PRAISE FOR J. B.

I0062426

J. B. Maxwell has taught me so much. I'm so grateful to have found his books.

— JOHN ARBUTUS CO-FOUNDER OF THE SEASONED GARDENER PA

Magnificent! Love the recipe aspect in each chapter!

— SUE MEADOWS

J. B. Maxwell delivers another helpful and useful book that we can all learn from and come back to time and time again.

— NICOLE TALA OWNER OF THE GARDENING STORE IN CENTRAL PA

NORTHEAST FORAGING FROM YOUR BACKYARD HOMESTEAD

NATIVE HERBALIST'S GUIDE TO IDENTIFYING 101 TASTY WILD EDIBLE FOODS

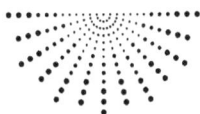

J. B. MAXWELL

To my loving supportive Wife, your constant warmth is like the morning sun.

© Copyright 2022 - All rights reserved.

The content contained within this book may not be reproduced, duplicated or transmitted without direct written permission from the author or the publisher.

Under no circumstances will any blame or legal responsibility be held against the publisher, or author, for any damages, reparation, or monetary loss due to the information contained within this book, either directly or indirectly.

Legal Notice:

This book is copyright protected. It is only for personal use. You cannot amend, distribute, sell, use, quote or paraphrase any part, or the content within this book, without the consent of the author or publisher.

Disclaimer Notice:

Please note the information contained within this document is for educational and entertainment purposes only. All effort has been executed to present accurate, up to date, reliable, complete information. No warranties of any kind are declared or implied. Readers acknowledge that the author is not engaged in the rendering of legal, financial, medical or professional advice. The content within this book has been derived from various sources. Please consult a licensed professional before attempting any techniques outlined in this book.

By reading this document, the reader agrees that under no circumstances is the author responsible for any losses, direct or indirect, that are incurred as a result of the use of the information contained within this document, including, but not limited to, errors, omissions, or inaccuracies.

CONTENTS

"Let thy food be thy medicine and medicine by thy food."

- Hippocrates

INTRODUCTION

We need plants. Plants are around us all of the time, but often we are not fully aware or appreciative of this. There is a term for this called "plant blindness." The basic definition of "plant blindness" is a term describing our tendency to be oblivious to the plants around us. Take, for example, wood. Think about how much wood we use in our surroundings, but it goes farther than that. All around the world people are connecting to plants by studying them, eating them, or finding beauty in them.

Plants keep us alive. The problem is that the world is getting harder and harder to live in—for both us and the plants. There is a lot of stress about the state of the world and what we can do as individuals, not only to help but also to find peace and purpose among this chaos. Everyday people are thrown statistics like this:

- The World Wildlife Fund (WWF) found that humans are using 1.6 times the amount of resources we have on earth.

- According to the University of Michigan (MIT) between 1990-2020, 75% of all agricultural crop biodiversity was lost.

The loss of species is not only in the animal kingdom, and the stress of the climate crisis can feel overwhelming to the point of doom for many. Foraging is a two fold solution in this case. Foragers are typically people who have formed a connection to the land, are conscious of the amount they take, and how to give back to it. Foraging for food, at least as a supplement to store bought, is a hobby that positively affects the environment and your personal footprint because it does not require clear cutting large acres of diverse land.

Among the stresses of the larger issues of the world, people are so busy with their lives that they don't get a chance to utilize the wilderness. People have been segregated from the outdoors and it's unnatural. Foraging is mentally stimulating in the research stage and it creates a deeper connection to nature. This common feeling of dissociating or living in a stimulation can be in part blamed on the disconnect that people have with the outside world. Humans are not meant to be blind consumers. They have always been part of the food chain and removing this natural habit and purpose is unfulfilling. While this sounds complicated and scary, we should just think of ourselves as way more simple. Like pets, sometimes people need mental stimulation that mimics what we are innately built for.

Not only are there mental reasons for adding foraging for food as a habit or hobby, there are physical reasons as well. Being out in nature is an excuse to increase exercise. For those who find it difficult to get the recommended amount of exercise, it feels different when there's a purpose that means more to them. Foraging is a passive exercise, but it is more in tune with the natural habits of humans and can counteract those immobilized by work. Foraging for food increases the diversity on your plate. According to the CDC (2022), only 6% of Americans eat enough fruit and vegetables. By starting a more intentional relationship with food, you will begin to see it in a more dynamic way. Once again, instead of this segregated exercise and diet culture, it is something that becomes integrated into your life naturally. It surpasses commercialized dieting and health; it opens up eating a more balanced diet that puts plants on the forefront.

As an average person trying to survive with food anxiety, money, lack of agency, and even feeling lost in the sea of information, many think that the solution has to be complex. The benefit of adding foraging into your tool belt is that you are able to feel like you're taking action, and you can feel less lost in the natural world around you. Being able to take control of food insecurity ensures that you feel comfort in knowing you can fend for yourself.

This book is set up as a dictionary about 100 plants you will be able to find in the Northeast. As you learn more about herbs and plants, you can eventually become self-sufficient, be more intentional with your health, and save money. As you slowly become more comfortable being in the world and interacting with the background, you will begin to see your tangible opportunities grow. Eventually you will feel comfortable using these plants like you would anything you pick up in the grocery store, and you might even start using them for natural healing. This book starts with the basics of foraging, then each chapter looks at one plant, specifically how to identify it, what it is used for, how to prepare it, and how to cook with it.

I grew up in a small farm town with no red lights in Maryland. I currently live in Pennsylvania with my wife and son. I am very family oriented, and they are a part of my life outdoors: hiking, nature, and gardening. Being outdoors brings peace to a world that is unpredictable. I have always loved the idea of fostering my own healthy, clean, organic food and producing enough food for the whole family and neighbors without having to rely 100% on the grocery store. I think teaching other people this knowledge about herbs and foraging wild herbs and plants is the first step.

There's something about producing your own food that gives it a taste unrivaled to anything anyone could ever buy. I absolutely love taking control of my own food and being more hands-on when it comes to providing for my family. I have been practicing homesteading principles on two different properties (one 0.75 acres, the other 1 acre) for a little over 8 years now. I have learned a lot throughout my life this way—from studying within the sphere of Western Medicine to traveling through the East and learning about Chinese Eastern Medicine and Ayurvedic practices. I have found great comfort and gratitude in having such knowledge about ancient wisdom. Because of all of the kindness of the spread of knowledge, I now live off-grid and wish to simply ensure that everyone else doing the same or similar feels safe and empowered as they do so.

I have acquired this knowledge by studying, listening, and doing hands-on experience like exploring the forest and mountains. I know that I am safe and protected no matter what happens, precisely because of studying wilderness survival and living it firsthand for over 10 years. This has led to a long and continued journey of knowledge expansion. Helping others achieve a healthier lifestyle and being able to be self-sufficient is my absolute passion because I know that it leads to so much freedom, joy, and better health. I want to be able to help others improve their relationship with Mother Nature.

If you're a food lover, you can expand beyond what is handed to you in the store and find what grows in the same cycles you live in as well. If you want to improve your health with foods and connect to the natural world, do it naturally. Herbs and plants have healing properties, and by eating a few of them daily, you are treating your body and your mind. It is only difficult at the start, and once you start putting plant names to their faces, the rest will be history.

1

FORAGING 101

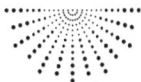

*F*oraging is the activity of collecting food from natural and wild sources. Picking berries off of a bush for a mid-afternoon snack when you are a child is a great example of foraging. Becoming a forager means adhering to the ethics of foraging and maintaining a relationship with the natural environment around you. The goal of sustainable and ethical foraging is to gather food from wild sources in a way that promotes the ecology of a given area. Foraging helps foragers to learn about the natural environment around them and how to work with those environments in ethical ways. Foragers understand that there is a give and take relationship within the natural world.

The ethics of foraging are incredibly important because they help you remain safe, promote ecological harmony, and practice legal foraging. In every region there are foragable plants that are in season and at risk. If you are outside of the Northeast, be sure to research both types of foragable plants when you start your foraging journey. The USDA plant map is a great online tool to help you discover what is available in your area.

This guide will help you through your foraging journey, but it is important to gather as much material as you can to learn how to identify all of the foragables in your area. Being able to accurately identify the foragables that you find can mean the difference between life and death. There are many dangerous and unsafe foragables that you are bound to find and without the proper knowledge, you could inadvertently harm yourself. Learning to identify plants will mean understanding what they look like, their leaf shapes, the fruits that they make, the general locations that they prefer, fragrance, and life cycle. Learn everything you can about what you may find so that you keep yourself safe.

When you are foraging, be sure to leave a majority of what you find behind. Depending on the abundance of a given foragable, it is important to only take up to a third of what you have found. This allows the natural life cycles and ecosystem to continue thriving without human interference. In addition to only taking a fraction of what you find of a particular foragable plant, be sure to leave the area that you are foraging in a good or even better condition than before you got there. Avoid drastically changing the landscape or disrupting the ecosystem to the best of your ability. This also means picking up any garbage you may have with you or cleaning up any that you find along your way.

In order to be the most ethical forager, it is important to know the laws surrounding foraging in your area. In the Northeast, state parks have specific requirements and restrictions around foraging. There are also laws in most Northeast states that prohibit foraging on private property. Be sure to look up the laws for where you want to forage and the zoning in the area you are in.

Making a Plan

When you go out for a day of foraging, be sure to have a plan that includes the following things:

- Know what you are looking for.
- Where you can find it.
- Take a picture of your new findings and consult a field guide to determine if it is safe to go back and forage for that plant. It is easy to get distracted and find something new to forage. Sticking to your foraging plan and minimizing distractions can mean avoiding potential danger by misidentifying a plant.
- In a foraging sense, clean areas are areas that are pollution and litter free. It's best to avoid foraging in urban and industrialized areas in order to ensure that you are finding safe plants to harvest and snack on.

You will also want to invest in foraging equipment to make the most of your excursions. The essential items that you will need when you head out on a foraging adventure are

- either a basket or a lightweight cross-body bag for collecting foragables
- a field guide
- zip-lock bags
- a hand lens for closer identification
- sturdy gardening or hiking boots
- gardening gloves
- a long sleeve shirt and pants

Some additional tools that you may consider bringing along, depending on your foraging plan for the day, are

- pruning shears or saws
- weeding or gardening knives
- a compact (pocket) knife
- a shovel
- a hoe
- scissors

Something to keep in mind as a forager is that you might not have interacted with these plants before. Even if you are someone who doesn't commonly have allergic reactions to things, allergies are the body's reaction to foreign bodies in the system. Whenever you are trying a new food, it's best not to feast on something you or your family has never interacted with before. Plants can contain things like latex that you might not even consider before you try. Even if it's not necessarily an allergic reaction, taking the time to see how your body reacts to these new foods means you stay safe and are able to decide whether you will be looking to forage for more in the future.

Foraging should not happen blindly. When you are out in nature, it is not a vacuum of the plant you want to forage and you. Each area comes with its specific situation, and it is up to the forager to know their skill level. This starts with their expertise level in being in nature in general. Those who have not spent time outdoors in wild spaces might find themselves greatly overwhelmed with the huge diversity of plants, as well as the exertion of being in nature—whether that be hiking, being in swampy areas, or even getting through thick brush. Beyond this, foraging isn't a task tackled in a day. Be sure to build up your skill level. Have respect for the wild, and solidify the respect you have for yourself in the wild.

WASHING YOUR FOOD

Everyone should wash their food, either when they bring them home or are about to cook with them. This includes produce, rice, and even cans before you open them. You don't know what has touched anything before you did. Even though most people clean their food when they cook, oftentimes no one has taught them how to clean your food effectively.

1. In a clean bin or bucket, add warm water. I do not recommend using your sink. Even if your sink is clean, it is still a place that usually holds bacteria, harsh chemicals, and more. A designated cleaning bin is more sanitary.
2. In this bin, add cold to hot water, depending on what you are cleaning. Leafy or delicate food should be done in cold water so it does not wilt. Anything more tough like roots or nuts should be cleaned in hot water.
3. Add about a cup of white or apple cider vinegar to the water. This won't make your food taste like vinegar since it will be rinsed, but it is a safe product to disinfect and clean your food.
4. Let it soak for 15 minutes.
5. Agitate the water and food, making sure any debris is loose. With hardier foods, you might even scrub.
6. Remove food from the bin. If there is visible dirt or the water is very dirty, rinse the bin and fill with water again. Submerge your food and give it a deep rinse, agitating and letting the remainder of the dirt rinse off.
7. Remove the food from the bin. Give it a rinse under cool water and place it on a dish towel.
8. Make sure your food is dry before storing or cooking with it.

2

IDENTIFYING YOUR PLANTS

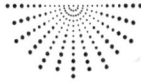

*T*here are a few ways to approach the identification process of a plant. It is really important that foragers become knowledgeable about the plants in their area before they go out in the field. Foragers who want to have success and confidence in foraging need to take the task seriously. While in many cases "inedible" does not necessarily mean "deadly," there are a plethora of unpleasant side effects to eating things that don't belong in the human body. Careless mistakes like this could really deter those from continuing down this path, but that does not need to be the case. There are lots of plants out there that you never knew were edible, so let's stick to those and figure out how they work.

If you have some background knowledge on plant names, it's important to know that there are often multiple nicknames or common names for a plant. This might cause confusion if you only know one variant of the name, so it's important to get familiar with its scientific name which is always more reliable. Another thing to remember when it comes to species and genus, there is often a species of a genus called the 'common' version, for example "common daisy" or Bellis perennis. It is common for plants that have the title 'common' dropped, which can be confusing when it comes to other variants in the genus. Before you forage something based on a broad name, you should identify its specific species for edibility. There are many flowers that look like daisies and are part of the daisy family, and they all range from what they are used for in food to how upset they could make your stomach.

In the last chapter we talked about the idea of a plan and why it's important to know what you are going for before you go out. In your research, you need to know what plants you are going for, where those plants can be found, and what time of year those plants are found before you need to worry about specifics on the plant itself. If you have a specific location that you want to check out for foragables, you can note the climate and create a list of plants you might hope to see. With this list, also familiarize yourself with plants you should avoid as well. This book tries to take note of some dangerous look-alikes to some specific plants, but that doesn't necessarily take into consideration location specific issues or plants like poison ivy or poison oak that can harm you and contaminate those plants around it with its oil.

As we start looking at the identification of plants, we need to gain an understanding of the vocabulary that is used. In this book there will be a mixture of common and scientific phrases. We want identification to be accurate, since scientific terms in some cases flag more specific characteristics, but gaining knowledge in this field does not require heavy research or even excessive page flipping. There is a learning process in understanding these specific characteristics that needs to go farther than 'leaf', but foragers should not feel like this book, or foraging in general, is not accessible to them. In each description, a reader is able to start with the general idea of a plant, then have a closer look at each of these details in a way that will allow them to apply their knowledge in the field.

When foragers are starting the identification process, the first thing they will typically notice is the form of the plant, then slowly work their way through the subcategories until they reach their answer. While there are some plants that can be more tricky, specifically when moving from small shrub and vine or bush herbaceous plants to large shrubs and small trees, subcategorizing them is the easiest and quickest way to narrow down plants. Below is a chart that represents how to move through the identification process.

Note: this book does not go through any mushroom identification.

IDENTIFICATION FLOW CHART

When identifying plants, there are multiple branching categories, the most used are family, genus, and species, which are the last three on the tree. When reading the scientific name, the first word is the genus, spelt with a capital, and the second word is the specifying species, spelt with a lower case; for example, the giant sunflower is called *Helianthus giganteus.*

FORM

Tree: Some plants can take the form of a tree or shrub. Single stem is woody—called a trunk, typically very hard or thick (large branches can look like multiple trunks). Typically over 10 feet is considered a tree.

Shrub (bush/ground cover/woody vines): Shrub is about 10 feet at its full height (this definition can vary) and has several woody or semi-woody stems. They can look dense like hedgewalls or sparse. *Some herbaceous plants can look "bushy" but are not actual "bushes."*

Bush: Bushes are shrubs on a scientific level. They are woody but tend to be shorter with branches that are not as densely packed.

Woody vines: Scientifically called *liana*, or sometimes called brambles. These can form shapes that look like shrubs but are *not* shrubs.

Ground cover: Any plant that covers the ground including grass. Short shrubs can be ground cover.

Thicket: An area of densely packed plants, including trees, shrubs, and woody plants. Sometimes called a brier.

Herbaceous: refers to plants that are not woody, but herbaceous plants can have very hardy stems, including some types of bamboo, tomatoes, etc.

Flower: Scientifically flowering herbaceous plants are categorized as **forbs**. Flowers are usually referred to as **annuals** (life cycle within the year and reseeds) and **perennials** (lasting though years).

Grass: Grasses are considered **graminoids**, a simple definition is blade like leaves and no flowers. Grasses do flower, but often they are non distinct.

Fern: Ferns reproduce by spores, not flowers or seeds. They produce a uncurling stem that expands into fronds (divided leaf)

Moss vs Lichen: Moss and lichen look very similar, but moss is a plant and lichen is both a plant and a fungus. Moss will have very small stems and leaves, and a lichen will appear more like a crusty layer of peeling paint, sometimes like little tufts of branches.

↓

Narrowing Characteristics (At least down to genus)
Tree: Coniferous (typically needles leaves) or Deciduous (typically herbaceous leaves) → Size (height, thickness of

trunk, and branches → Characteristics of leaves/needles (length, color, shape, how many, stem, pattern) → Fruit (size, hardness, color, texture, time of year) and Flower (size, color, shape, single or clustered, time of year)→ Seeds or nuts (size, hardness, color, texture, time of year) → bark (color, texture)

Shrub: Size (height, thickness of trunk, and branches) → Characteristics of leaves/needles (length, color, shape, how many, stem, pattern) → Fruit (size, hardness, color, texture, time of year) and Flower (size, color, shape, single or clustered, time of year)→ Seeds or nuts (size, hardness, color, texture, time of year) → bark (color, texture)

Flower and grass: Flower (size, color, shape, single or clustered, time of year) → Size (height and width) and Leaves (length, color, shape, how many, stem, pattern) and Stem (color, shape, how many, branching, texture) → Fruit (size, hardness, color, texture, time of year) → Seeds or nuts (size, hardness, color, texture, time of year)

Fern, Moss, Lichen: : Size (height and width) → Leaves (length, color, shape, how many, stem, pattern) →Stem (color, shape, how many, branching, texture)

↓

Forgeability and specifying species

Age of plant: The age of the plant from tree to flower can affect these characteristics.

Gender: Some plants have genders, this is especially notable in trees. This means that not all trees produce fruit. This is especially frustrating in urban environments where female trees, the ones with fruit, are not planted because they are more messy.

Time of year: Time of year affects when plants are producing foragable parts, but it can also help determine look-alikes from each other. Often plants that look similar are not flowering or producing at the same time, even from the same genus.

Climate/habitat: For plants you are actively seeking out, you already know where they like to grow, but for identifying a plant you have stumbled upon, checking the growing habits can eliminate many options, including plants of the same genus. This can include larger **geography** of the world, local **ecosystems and surroundings**, or **growing conditions** such as soil and sun.

Flower and leaves: Flowers and leaves can help narrow down to a family or genus, but a closer look at their biology can give a closer look at the subtle differences between similar flowers. See next part.

Leaf Biology

Start with the most obvious characteristics of the leaves

- Color: summer and fall
- Size: length and width
- Texture: stiff, limp, hairy, waxy, thick, wrinkled, or folded
- Shape: silhouette, lobed, and edges

Each leaf will have

- **Base:** The base of the leaf is the bottom edge.
- **Margins:** The margins are the distance between the veins and the edge of the leaf.
- **Tip:** The tip is the point of the leaf, there may be multiple tips.
- **Veins:** The veins are the small structures that run along the leaves to provide water and nutrients to the plant.
- **Midrib:** The midrib is the center vein on the leaf.
- Some leaves have **petioles** as well, which is the **stem** that connects the leaf to the bush, tree, or flower it is a part of.

Identify the shape of the leaf arrangements.

- If the leaf is a single structure, it is a **simple leaf**.
- If there are multiple smaller leaves that stem from the same place, you are looking at a **compound leaf.**
- The most common leaf arrangements that you will find while foraging are parallel, dichotomous, palmate, and pinnate leaves.
- **Parallel** leaves can be both simple and compound structures, where all of the veins are parallel to each other.
- **Dichotomous** leaves form a Y shape and are often part of a group of leaves.
- **Palmate** leaves resemble an open hand. Maple leaves in the Northeast are a perfect example of a palmate leaf arrangement. Palmate leaves have veins and a midrib that starts from a common point and extends outward through the leaf.
- **Pinnate** leaves are smaller leaves that extend from one central stalk or axis.

Identifying a plant's **branching patterns** can also help you to quickly identify the foragable that you are looking for.

- Plants that have **opposite branching** are those whose leaves grow on opposite sides of the stalk or axis. Mint is a great example of an opposite branching plant.
- **Alternate branching** plants have alternating leaf patterns, where the leaves connect to the axis on alternating sides one at a time. These two types of branching patterns are the most common in the Northeast.

Flower Biology

You may also be foraging for **flowers** or blooming foragables.

First, you will probably notice these characteristics of a flower

- Color: single or multiple
- Size: unnoticeable (grass), small (babies breath), medium (daisy), big (rose), or large (sunflower)

The next thing you will probably notice is the **Inflorescence.** This is a cluster of flowers. This can get complicated and specific, but in this book it will usually refer to the clusters in descriptive words, like 'floret', for ease. The shapes of the inflorescence/floretes are categorized as spike, catkin, racemose, umbel, head, and more.

- **Shape:** Flower shape, or petal arrangement, can often best be described in reference to common or well known flowers.
- Many flowers are simple, 5 petal flowers that fan out flat from the edge of the center and are described as **star-shaped**.
- **Trumpet or funnel** flowers usually have petals that fan out at the ends or cup, but tapper down in the center like a funnel; for example, petunias.
- **Cup** shaped flowers have petals that don't spread out flat. Instead, they create a cup or bowl shape, like tulips or buttercups.
- **Global** flowers, like dandelions, have a ball shape.
- **Bell shaped** flowers typically are shaped like bells **and** hang from their plant.
- **Tube** flowers can be single petals or multiple that form a narrow tube.

- **Rosette** flowers that have petals that create a dense overlapping, usually in a circular pattern.
- **Flat, fanned rosette** flowers are like daisies that are flat, like a star shaped flower, but have more petals.
- **Non-symmetrical** flowers are flowers that might have a petal like a tongue or hood.

Like a solitary flower or a floret, the individual flowers can then be observed by

- Number of flower petals
- Shape of the petals
- Brackets (joint of flower and stem)
- Center (stigma, stamen, pistols, ovary, etc.)
- Sepals (sometimes leaf like, sometimes petal like, this is the outside of the flower as its blooming)

3

LIVING IN TUNE WITH NATURE'S CYCLES

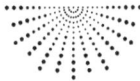

S **pring Foraging**
Spring is a time for growth, renewal, and creativity. Living cyclically with nature during the spring can leave you feeling full of energy and confidence. You will want to learn more and explore after the winter season. This might be the first time for many to get back out in nature. Take action, plant, and forage during the spring to capitalize on these feelings.

When it comes to spring foraging, the flavors are fresh and crisp with all the new growth. A lot of your foraging options are early shoots and leaves. If you are a new forager, this can be difficult because shoots often do not have the most distinguishing characteristics of the plant yet. It can help if you know where you have seen these plants grow in previous years or have spent time observing them in different stages.

Foraging in spring is often a little time sensitive. Flowers can bloom for short periods of time, and young shoots are only edible when they are tender. It is a good idea to preserve what you can because, in many cases, they won't be there for the rest of the year.

Plants found in spring

- Asparagus (Shoots)
- Amaranth (leaves)
- Basswood (flowers)
- Birch (twigs, sap)
- Burdock (stems)
- Cattails (shoots)
- Chickweed (greens)
- Chicory (leaves)
- Cleavers (stems)
- Cow Parsnip (stems, leaves, and flowers)
- Curly dock (leaves)
- Dandelion (all)
- Daisy (leaves and flowers)
- Eastern Redbud (flowers and buds)
- Epazote (leaves)
- Evening Primrose (root and leaves)
- False Solomon's Seal (shoots)
- Fiddleheads (stems)
- Field Garlic (leaves)
- Garlic mustard (leaves and stems)
- Ginkgo (leaf)
- Glass weed (stems)
- Goutweed (leaves)
- Groundnut (root)
- Henbit (flowers, leaves, and stems)
- Honewort (leaves)
- Japanese Knotweed (leaves, stem)
- Lady's thumb (leaves)
- Lambs quarters (leaves)
- Leeks (root and leaf)

- Maple (spring)
- Milkweed (shoots)
- Mint (leaves)
- Mugwort (leaves)
- Mulberry (fruit)
- Nettles (leaves)
- Plantain (leaf)
- Pokeweed (leaves)
- Purple Dead Nettles (leaves)
- Quickweed (leaves, flowers, and stems)
- Red cover (leaves and flowers)
- Sassafras (leaves and twigs)
- Sweet clover (leaves)
- Violet (flowers)
- Watercress (leaves)
- Wild lettuce (leaves)

Summer Foraging

Spring comes quickly, and before you know it the world is just buzzing with life again. I find one week the trees are bare, and suddenly overnight, they are lush and green again. This is a time of energy and prosperity. The summer is a time that overlaps with harvest after harvest. For those learning to forage, it is a great time to get to know plants at their fullest. The plants you might already be familiar with are great learning blocks to observe closer. Take note of things that you may not have thought about. Expanding your knowledge of what is already familiar can be easier than introducing yourself to completely foreign plants without being able to reference the characteristics to something you know. Foraging in the summer has a plethora of options. It can be over-whelming too. Make a plan to go out and observe without foraging and get used to your local plants. You don't need to know everything, but getting a foot in the door can help.

Even though it's a lot at once, this is the time to think ahead too. Try to notice where winter, or even spring, foraging can happen? Plants are the most recognizable now, so take notes or map out where these plants can be found. Since the time is prosperous, divide your time between foraging for fresh food and foraging for preservable plants. Build your inventory of dried, jam or jelly, frozen, pickled, and canned foods.

Plants found in summer

- Amaranth (leaves, seeds)
- Black cherry (fruit)
- Beech Plum (fruit)
- Beebalm (leaves)
- Birch (twigs)
- Black cherry (fruit)
- Blueberry (fruit)
- Brambles (fruit)
- Cattails (flower and pollen)
- Chicory (root)
- Common Elderberry (berry)
- Strawberry (fruit)
- Cherry Dogwood (fruit)
- Curly dock (leaves)
- Current (fruit)
- Dandelion (all)
- Daylily (flowers and root)
- Enchanter's nightshade (fruit)
- Epazote (leaves)
- Field Garlic (leaves and flower)
- Garlic mustard (leaves and stems)
- Ginkgo (leaf)

- Glass weed (stems)
- Goldenrod (leaves and flower)
- Goutweed (leaves)
- Ground Cherry (fruit)
- Grape (fruit)
- Juneberry (fruit)
- Lilac (flower)
- Mint (leaves)
- Mugwort (leaves)
- Mulberry (fruit)
- Mustard (leaves and flowers)
- Nasturtium (leaves and flower)
- Daisy (leaves and flowers)
- Quickweed (leaves and stems)
- Pepperweed (leaves)
- Plantain (leaf)
- Red cover (leaves and flowers)
- Rose (flower)
- Quickweed (leaves, flowers, and stems)
- Queen Anne's Lace (fruit and flowers)
- Wild bean (seed)
- wild ginger (root)
- Watercress (leaves)
- Wild lettuce (sap)

Fall Foraging

The fall is a time characterized by slowing down. There is something in the air as fall comes close that feels like the ending. This is a time of closure, wrapping things up, and bringing comfort. There is a sort of before and after in the fall that comes with the first frost. Most herbaceous plants die at this point, most fruit goes bad, and those should be a priority in foraging before any heartier plants. While the fall is the time of closing, it is also a time to harvest for many crops. Those who have their own garden and grow squash and more are finally seeing the fruits of their labor.

Because winter is coming, the forager should take advantage of any fresh food that they can get while they still can. Stocking up on preservatives, especially the late summer into fall fruit, is a good idea. If you have a garden or backyard plants, they might need to be protected over the winter.

- Ameranth (leaves and seeds)
- Apples (fruit)
- Barberry (fruit)
- Beebalm (leaves)
- Birch (twigs)
- Black cherry (fruit)
- Black walnut (nut)
- Butternut (nut)
- Cattails (roots)
- Common Mallow (root)
- Cranberry (fruit)
- Curly dock (leaves and seeds)
- Dandelion (all)
- Evening Primrose (root and leaves)
- Epazote (leaves)
- False Solomon's Seal (berries)
- Field Garlic (leaves)
- Fragrant sumac (berries)
- Ginkgo (leaf and nut)
- Glass weed (stems)
- Goldenrod (leaves and flower)
- Goutweed (leaves)
- Ground Cherry (fruit)
- Grape (fruit)
- Hawthorn (fruit)
- Highbush cranberry (fruit)

- Hickory (nut)
- Honewort (root)
- Groundnut (root)
- Jerusalem artichoke (root)
- Juniper (berry)
- Lambs quarters (seeds)
- Lotus (root and seed)
- Mayapple (fruit)
- Mint (leaves)
- Mustard (leaves, flowers, and seeds)
- Oak (nuts)
- Parsnip (root)
- Pawpaw (fruit)
- Peach (fruit)
- Pear (fruit)
- Pepperweed (seeds)
- Pineapple weed (fruit)
- Plantain (leaf)
- Rose (fruit)
- Sassafras (root)
- Thistle (root)
- Autumn Olive (berry)
- Queen Anne's Lace (root)
- Watercress (leaves)
- Wild Potato (root)

Winter Foraging

Winter is a time of rest, peace, and often family. For those who get snow, winter can be a very isolating time, especially if you find an escape in the outside world. On the nicer days it's a good idea to try to get outside and enjoy the sun as much as you can. Luckily, you can forage in the winter.

Foraging in the winter is obviously the most complicated, even more so in places with snow. For some, this might be a welcome challenge. Most foraging in the winter is done by finding any nuts or fruit that wasn't harvested in the fall. If the ground is not frozen, roots are a great option, as well as some cold resistant greens.

Here are some plants to forage during the winter:

- Common Mallow (roots)
- Curly Dock (leaves)
- Dandelion (all, but mostly root)
- Evening Primrose (root and leaves)
- Fragrant sumac (berries)
- Highbush cranberry (fruit)

4
AMARANTH

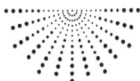

*maranth is a name for a genus of plants under the scientific name *Amaranthus*, also known as 'Pig-weed.' The most commonly foraged species of amaranth is *A. retroflexus* and *A. hybridus*, but many of the genus can be eaten. This plant should be used with caution as it pulls high amounts of nitrogen from the ground.

IDENTIFICATION OF THE PLANT

The two species noted above get as tall as 7 feet with a single erect stem, but other plants in the genus range from 3-10 feet tall. The plants are usually an average green and may have a red hue.

The characteristics of the leaves are as follows: simple, ovate alternating leaves with a slight waved edge on some species, some with more prominent points, and some that are completely smooth. *A. retroflexus* leaves reach up to 4 inches long and 3 inches wide. *A. hybridus* can be up to 6 inches long and 2 inches wide.

The flower is very small and densely packed into a spike.

A. retroflexus has a more weedy look, with a green flower that takes a spike shape at the top of the plant that reaches up to eight inches, is fairly narrow, and typically seen late summer to fall. This plant is more ornamental and goes bright red like some garden varieties. The spike might have slightly longer offshoots, making it look like a bunch of caterpillars. However, it has less of a droop than some of the other more decorative species.

Where to gather it

Amaranth is a common weed. As such, it likes most places, including areas that are overgrown, meaning places where the ground has been disturbed at some point. It can be found in spring, summer, and fall. It thrives in temperate climates, which makes it perfect for the Northeast.

How to gather it

In the spring, gather the greens. Once it is established, it will become woody. Seeds can be harvested from the flower spike, but be sure to time this right because the seeds have to develop before they drop to the ground. You might place a sort of mat on the ground to collect the seeds by shaking the plant or by gently cutting the spike off completely.

How to cook with it

Edible: the seeds of the plant are packed with protein. They can be eaten raw or you can roast them, use them as a porridge, popcorn, ground into flour, etc.

The leaves can be eaten raw like regular leafy greens or you can boil them, similar to spinach, or dry them and use them as a herb. They have a hearty flavor to them mixed with a hint of sweetness.

AMARANTH PORRIDGE
    Time: 25 minutes
    Serving Size: 4 servings

Ingredients:

- 1 cup of Amaranth seeds
- 2 cups water
- 1 tsp honey for taste
- 1 pinch salt for taste
- optional: other warming spices like nutmeg and cinnamon

Instructions:

1. Bring water and salt to a boil.
2. Add Amaranth seeds and reduce heat to a simmer.
3. Cover the pot and let simmer for about 20 minutes so the seeds absorb the water.
4. Take the hot cereal off of the heat.
5. Add honey and spices of your choice (optional).

Author: Pigweed is desirable because it provides vitamin A and C as well as a balance of other minerals and daily nutrient needs. It has been used in medicine for soothing common symptoms of ailments. It is also used as a natural yellow and green dye.

5
APPLES

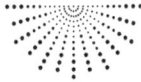

*T*here are many kinds of apples, all genetically selected and bred based on people's specific preferences. This identification should help you determine that the tree belongs to the genus *Malus*, but apple trees are particularly hard to identify down to the specific species. This is because the seeds of an apple are unpredictable, using genetic code completely different from the apple that it came from. This occurred as a result of settlers using apples for drinks, not typically for food, and not caring to grow apples from grafting, which is the only way to ensure the kind and quality of apples. Finding an apple in the wild is luck of the draw when it comes to quality. The difference between an apple and a **crabapple** is the size of the fruit. An apple that is four inches across or smaller is considered a crabapple.

IDENTIFICATION OF THE PLANT

The trees can grow up to 40 feet by a huge range, the smallest being more shrub-like than a tree at 10 feet tall. They typically have a solo trunk and a mushroom-like top.

The trunk is light gray to brown that cracks and peels from a smooth surface as it matures. The leaves are up to four inches long. They are a simple shape, slightly thick in texture, smooth to the touch, and lighter green.

The flower of the apple tree has five petals, which are typically white but can be pink or even red. These grow on long stamens, but can be bunched together on the tree, giving the blooming tree a dense look.

The fruit of an apple ranges quite a bit in between species. Crabapples are typically smaller. Crop apples can get as large as five inches wide. They are round and range in color from red, to yellow, green and more. When cut in half, the apple will have a star-like pattern at the core that contains the seeds.

Where to gather it

Apples bloom in the spring and are ready to pick in the fall. They like areas that have a lot of sun and slightly shelter, possibly from surrounding trees. They like well-drained soil with just enough moisture.

How to gather it

Apple blossoms are edible, but too much can be poisonous. Forage at your own discretion as picking blossoms can mean not letting an apple grow in its place. They would be plucked off the tree in the spring. Apples are ready to eat in the fall. The stem will be brown, making it easy to twist off the tree. You can also taste test if you know the species to see if the apple has reached its desired flavor.

How to cook with it

Edible: Yes.

Apples in nature are the same as apples from the store. Depending on the flavor of the apples, you should determine the best way to use them—whether raw or in pies, drinks, or sauces. Apple blossoms can be used as a floral in spring salads.

. . .

Apple Crisp Recipe
Time: 1 hour
Serving Size: 6 servings
Ingredients:

- 2 cups peeled and chopped apples
- 1 cup flour
- 1 cup white sugar
- 1/2 cup softened butter
- 1 tsp lemon juice
- ¾ cup rolled oats
- a pinch of salt
- warming spices like nutmeg and cinnamon

Instructions:

1. Preheat the oven to 350 °F.
2. In a bowl, mix the apples, spices, lemon juice, and sugar.
3. In a separate bowl, add the oats, flour, salt, and butter.
4. Cut the butter into the dry ingredients until the pieces are the size of a pea.
5. In a buttered baking sheet, add the apple mixture.
6. On top of the apples, layer the dry mixture evenly.
7. Bake for about 45 minutes.

Author: Apples have a fairly good shelf life, even as a raw fruit. It's possible to pick more than a day's worth and not waste anything. Eating a freshly picked apple from a tree and enjoying the crunch of it on a sunny autumn day is one of the most amazing feelings.

The leaves and tender twigs are an agreeable food to many domestic animals, like cows, horses, sheep, and goats; the fruit is sought after by the first, as well as by the hog. Thus, there appears to have existed a natural alliance between these animals and this tree. The fruit of the Crab in the forests of France is said to be a great resource for the wild-boar.

6
ASPARAGUS

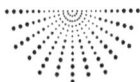

*I*f you've never seen an asparagus—scientific name *Asparagus officinalis*—growing from the ground, you might be caught off guard to see one naturally. Young asparagus shoots come out of the ground, and it looks like someone just stuck a piece of asparagus in the ground.

IDENTIFICATION OF THE PLANT

Wild asparagus looks as described above in the spring: a stalk up to 12 inches long, erect, with a scaly pointed top. As it matures, it gets up to six feet tall and very thin and spindly. It looks like an herbaceous tree by the fall if it hasn't been harvested, with lots of thin branches and thin, dill-like needles.

The flower is a pale yellow that hangs off the stem in opposite pairs with about a 2 inch stem. The flower has 6 sepals that create a long bell shape and a small golden pistils (interior/center).

The fruit of the asparagus is a bright red, blueberry sized

berry that hangs off the stem in opposite pairs with about a 2 inch stem. It contains round black seeds that are toxic.

Where to gather it

You can find asparagus in well drained areas that have been disturbed. If you are able to spot the larger, mature plant out of season, take note and come back in the spring. You might be able to notice last year's remains to help identify their spot.

How to gather it

Carefully take a sharp pair of scissors or a knife and cut the young shot an inch or so from the ground. You might be able to come back and find more shoots as well.

How to cook with it

Edible: Cook young shoots for about 15 minutes in whatever recipe you choose. They can be boiled, steamed, or roasted. Do not eat mature plants as they can be poisonous.

ROASTED ASPARAGUS

Time: 20 minutes

Serving Size: 4 servings

Ingredients:

- 10-15 Asparagus spears
- 5 tbsps Olive oil
- garlic, adjust to preference
- 1 pinch salt and pepper for taste
- 1 pinch sesame seeds

Instructions:

1. Preheat the oven to 425 °F.
2. Lay the asparagus spears on a tray flat.

3. In a bowl, mix everything else, then drizzle over the asparagus, but make sure they are covered.
4. Cook for up to 15 minutes.

Author: Asparagus is high in fiber and antioxidants and are commonly used for detoxing. White asparagus is found in places where the young shoots are covered to keep them away from sunlight for better flavor; however, it has less nutrients. The seeds and roots of asparagus have been used in medicine, but there has been no scientific evidence for it. I prefer to boil them freshly picked and eat them with salt, lemon, and hollandaise sauce on the side.

7

BASSWOOD

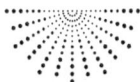

*merican basswood can also be called Liden or *Tilia americana*.

IDENTIFICATION OF THE PLANT

This plant is a tree with leaves that grow up to 120 feet tall and 5 feet wide, making it rather huge when it has reached maturity. As long as the tree is under 100 years old, it will probably seed. The bark is gray to brown and smooth. The newer growth on the wigs are green with a reddish hue.

The leaves are heart-shaped with a saw-like edge, and more of a darker green with a slight gloss to it. The leaves have a pain vein down the center with parallel straight veins that spread out of it. The young buds of the flower have got a red hue to it.

The flowers hang from long stems that branch off to more flowers. They are small and yellow, and they start round and open to long, fanned stamens. The fruit is small, round, and hard. It starts light green, then turns brown.

Where to gather it

The buds are some of the most desirable parts of the tree, which come out in very early spring. The flowers bloom in late spring and into early summer.

How to gather it

If you can find a tree with branches that are low hanging enough in the spring, you can pick or cut off the buds. The leaves are always edible, but taste better when they are younger. Similarly with the flowers, they can be plucked off of low hanging branches when they are in bloom.

How to cook with it

Edible: Yes, all parts of the tree can be safely eaten, but the buds and the flowers are the only parts typically eaten for enjoyment.

The buds are eaten raw or dressed up in salad. The leaves are edible all year, but are best as buds. If you like the mature leaves, they can be used up until the fall when they fall from the tree. The flowers can be dried for tea. The tree can also be tapped to make a syrup.

LINDEN TEA

Time: 2-20 minutes
Serving Size: 1 serving
Ingredients:

- a few tsps dry linden flower
- 1 cup water
- honey for taste

Instructions:

1. Put water on the heat and remove just before boiling.
2. Add leaves to your tea pot or cup.

3. Let steep for 2-20 minutes.
4. Remove tea leaves by removing the bag or strain.
5. Serve hot.

Author: This tree is a favorite because not many people can eat the leaves off of a tree. While the leaves are preferred when they are younger, it can be a funny party trick to start eating a leaf right off a tree.

The first time I ever had it was in my grandma's house when I was a kid. She prepared this delicious salad, and instead of lettuce, she used basswood leaves. It was so tasty and quickly became one of my favorite dishes at her house.

8

BARBERRY

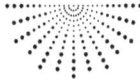

*B*arberry is a name for a genus of plants scientifically named *Berberis*. The berries from the barberry plant can be picked from many of the species, but there is a strong preference for specific species, and there is more evidence about safety with these common species. The barberry is invasive, and in some places in the United States is actually illegal to plant. If you are foraging in your backyard, it is a good idea to keep your bush from spreading and taking over other plants.

Other things to remember when foraging the barberry shrub is that they can be a home to ticks, thorns, and rust. Rust is an orange colored mold that can infect other plants. Ticks are bugs that can carry lyme disease. They are no joke, unlike other bugs that are annoying, these bugs are not harmless and can cause long term health issues. Be safe. When you are foraging, make sure to wear thick, long pants and high boots, some even duct tape the seam of their boot and their pants to make sure that the ticks have no way in. Wear gloves and check yourself and any pets you brought along with you for ticks if you come in contact with one.

The species of barberry that this chapter will cover is *B. Vulgaris.*

IDENTIFICATION OF THE PLANT

The bush can grow up to 10 feet, but it is common to see them smaller. The shrub is not necessarily dense, but the long, spindly branches can create quite a cover. They are thin and reddish yellow when they are young, turning gray as they mature. The branches are covered in long thorns. The leaves are alternate, but they seem to come out of the branches crowded in rows of about 5, creating a circular fan look. The leaves are about 2 inches long and have an oval shape that is rounded on the ends. They have a prominent margin and slightly fold into it. It is also slightly toothed on the bottom. The color can range from a yellow green with a very faint red rim around the edge to a dark, attractive red in the fall.

The flowers bloom in the spring and droop in triangle florets, which is the shape and size of a regular grape cluster. The individual flowers are yellow and form a round cup look, opening wide to a round flat, pale green disk on a stigma. The fruit droops in clusters of about 10. They are a bright red, although more orange than purple. The individual fruit is about half an inch long and oblong like a stretched out grape. Similar in appearance to the cranberry.

Where to gather it

The shrub is invasive, which means that it is typically fairly adaptable, although it prefers more open areas. Any disturbed or typically weedy area is a good place to start. The berries are ready to pick in the fall. The red color of the berries is a good indicator they are ripe.

How to gather it

As noted above, it's important to follow more serious

safety precautions when harvesting from the barberry. When the berries are in season, you can pick the berries off of the bush. If the area is infested with barberry, you might have a harder time getting through the thicket.

How to cook with it

Edible: Yes, the ripe berries can be eaten, but nothing else.

The berries can be used in jams and other preservatives, but also has a place in savory dishes. It's especially popular in Persian cooking.

BARBERRY RICE

Time: 40 minutes
Serving Size: 4 servings
Ingredients:

- 1 cup rice (basmati recommended)
- 1 tbsp unflavored greek yogurt
- 1 pinch saffron
- ⅓ cup barberry
- 1 tsp sugar or honey
- 1 pinch salt and pepper, onion, garlic (other spices optional)
- 1 tbsp oil

Instructions:

1. Clean and wash the rice and berries.
2. Put the barberries in a heat safe container and cover with 1 tbsp of boiling water.
3. Repeat in a separate bowl with saffron.
4. Wash the rice and add it to a pot with water and salt.
5. Bring it to a boil on high heat.

6. Once the rice has hit a boil, add the barberries.
7. Cover, turn down to simmer.
8. About 15 minutes later, when the rice is ready, take off the heat, keep the lid on, and let it sit for 5 minutes.
9. Fluff after 5 minutes, add pepper, saffron, and saffron water, but don't overmix.
10. Serve with greek yogurt, add in spices based on your own preference.

Author: The barberry plant is poisonous since it contains a chemical called berberine. This chemical, while poisonous as a food, is used as a medicine against infection, especially orally. Unless you have the proper research on how to use this plant as a medicine, you should stick to a professional provider's advice, but if you are getting into the multiple uses of a plant, this is something to consider.

9
BEACH PLUM

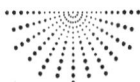

There are a few varieties of the beach plum, but this species is called *Prunus maritime*.

IDENTIFICATION OF THE PLANT

This plant grows up to 10 feet tall but is considered a shrub. The branches are slightly hairy when young and produce alternating leaves that are three inches long and simple with a slightly jagged edge.

When the plant flowers, it has a few flowers per cluster, they are only about half an inch wide and are white in a star shape. The fruit, which is round and rather small (i.e., slightly larger than a cherry), is the most important part of the plant. The color of the fruit varies, just like normal plums, from dark red, blue, or yellow.

Where to gather it

Beach plums live up to their name because they like to grow in sandy areas. The fruit is ripe in late summer.

How to gather it

A beach plum can be enjoyed by plucking it from the tree.

You can taste test it to see if the ripeness of the fruit is to your liking.

How to cook with it

Edible: Yes, you can eat the fruit.

It is recommended to use these plums in jams or the like, desserts, or turn them into prunes.

BEACH PLUM JELLY

This recipe makes about 8 jars of jelly.

Time: about 30 minutes

Ingredients:

- 8 cups beach plums
- 1 cup of water
- 6 cups sugar
- 3 oz liquid pectin
- optional: warming spices like nutmeg and cinnamon

Instructions:

1. In a large pot, bring the beach plums and water to a boil.
2. Cook until the fruit is soft.
3. Strain juice from any flesh or pits.
4. Pour 4 cups of this juice back into a clean pot with sugar.
5. Sugar should dissolve and come to a boil.
6. Add pectin and bring to a boil again.
7. Remove from heat, start adding it to jam jars that have been sterilized.
8. Don't wait until cool because it will turn solid and be harder to jar.

9. Seal the jar and store for up to a year, make sure to label.

Note: Once the jars are sealed, it's recommended to take the ring off of the jar. This ring is only used to keep the top on when initially sealing. If it is kept on the jar and if the seal breaks, the ring can reseal the lid, making the jar contaminated and rotten without you knowing.

Author: The pit is rather large for this fruit and some might find this recipe too much work because of that. The pit should not be ingested because it contains chemicals poisonous to humans.

The first time I tried this plum straight from the tree it was so acidic that my entire face flexed. I told my friends that this was the best plums they'd ever have and so they'd bite into them and feel the same sensation. I loved playing this game as a kid with my friends, sister, and family.

BEE BALM/BERGAMOT

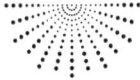

*B*ee balms and Bergamot are part of the genus *Monarda*. Specifically, wild bergamont is called *M. fisulosa*. This plant is actually part of the mint family.

IDENTIFICATION OF THE PLANT

The plant is small and creeps across the ground, but can grow up to three feet tall. The flowers bloom on the top of stems, but they typically grow in clusters. The color of this species of flower is a light purple. They are not cultivated for their ornamental use, unlike others from this genus. The flowers are less flashy, opening with thin 10-14 petals and fanned stamen.

The leaves are opposite and ovulate. The color varies a bit from light to dark green.

Where to gather it

This plant grows in rich soil but likes well drained areas. They like sunlight with protection, and, as such, they are found near openings of trees. This plant blooms in the summer and into the fall.

How to gather it

You can harvest the leaves and flowers off the main plant or cut the whole thing and garble it inside.

How to cook with it

Edible: The leaves are edible raw or cooked. They can be used in salad or boiled in cooked greens. The flowers are used in teas for floral taste and can be dried.

BEE BALM BREAD

This recipe makes two small loaves.

Time: 2 hours

Ingredients:

- 1 cup 1/2 warm water
- 1 pack yeast
- 4 cups flour
- 1 cup bee balm petals
- 1 egg
- 1 tbsp softened honey
- 3 tbsps butter

Instructions:

1. Activate yeast in a bowl with ½ cup warm water, let sit until it creates a bit of a foam (around 30 minutes in a warm room).
2. After the yeast has activated, add the rest of the water, butter, and honey, then mix.
3. Add flour and bee balm. Stir together well.
4. Form into a ball and cover. Let it rise for 1 hour.
5. Take a ball of risen dough and knead for 10 minutes.

6. Bread should be stretchy, so that it does not rip right away.
7. Preheat the oven to 400 °F.
8. Let sit for 20-30 minutes, while preheating in a baking pan or split into 2 pans.
9. Egg wash a beaten egg gently onto the top of the dough.
10. Bake for 30-40 minutes or until bread looks golden on the top.

Author: This plant is used and noted for a multitude of reasons, starting with attracting hummingbirds, due to its pleasant smell. It is also used in herbal medicine for relieving common cold symptoms and more.

BIRCH

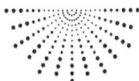

*B*lack or sweet birch, also known as *Betula lenta*.

IDENTIFICATION OF THE PLANT

The tree typically only grows to about 30 feet but has the possibility of growing up to 100 feet in the right habitat. The bark is similar in texture to the typical peeling, like the thin bark that is associated with birch trees, but it is a dark gray. If scratched, the young branches emit a minty smell. As the tree matures, it leaves the thin bark behind and becomes thick and cracked. The leaves alternate; they grow up to six inches long. They are an elongated, simple shape, with a main vein down the center and parallel veins moving out and upward straight from it. The edges are very finely jagged. The color is green to yellow in the fall.

The flowers are drooping two inch catkins. They are thin, small, ball-like flowers. They are yellowish green to a burnt yellow.

Where to gather it

The trees can be found commonly in forests, they prefer well drained soil. It can be harvested any time of year, which is especially useful and tasty in winter. The sap is best harvested in the early spring when the sugars are flowing.

How to gather it

It is best to avoid eating the inner bark of a tree because it can cause a lot of damage to the tree. Young twigs can be cut off the tree. If there are any roots that are at ground level, they can be sawed and harvested.

How to cook with it

Edible: Yes, twigs, inner bark, root bark, and sap.

The inner bark is only used in emergency situations. It can be dried or boiled. When dried, it can be grinded and made into a flour. Tea can be made from steeping twigs and inner bark. Sap needs to be boiled into a syrup. All parts need to be cooked first.

BIRCH TWIG TEA

Time: 45 minutes

Serving Size: 4 servings

Ingredients:

- 1 cup Birch twigs

Instructions:

1. The birch twigs should be fairly young and small. Cut them into small pieces, about an inch long.
2. Heat the oven to about 325 °F.
3. On a tray lay the pieces out flat and evenly.
4. Roast twigs for about half an hour.
5. Let rest until cool.

6. In a mortar and pestle or other grinding device, gently grind the twigs. Do not over grind, this is a tea, not a coffee. If it is over ground it will be harder to strain leaves after steeping.
7. Store dry tea in a dry and cool area.
8. Use one spoon of tea leaf for 1 cup of hot water to make tea.
9. Because of the roasting and grinding process, you only need to steep for up to 5 minutes, unlike other wood or root teas that need to be boiled or steeped longer.

Author: This is not a plant that needs to be at the top of the foragers list. Preferably, it should be used by those who may want to try some more advanced methods and plants for foraging.

It was in Norway. This tall bearded man with long blond hair took me into the woods. He brought a bottle, a string, and a metal straw and told me that I'd be about to drink the best water of my life. He stopped at a Birch tree and dug a little hole to put the metal straw inside, attached the bottle, and told me to prepare myself mentally for this elixir. We kept walking. After 1 hour we went back to the same tree. The 1-liter bottle was filled with this birch water, called birch sap. I took a sip of it and felt my body healing. It's something magical that everybody needs to try at least once in their lives. Don't forget to be respectful towards nature. Don't drain the tree because, ultimately, this sap is the tree's blood.

12

BLACK CHERRY

⁂

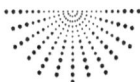

*T*he black cherry is scientifically named *Prunus serotina*.

IDENTIFICATION OF THE PLANT

The tree grows up to 70 feet. The bark of this tree is fairly dark bark that is smooth, but cracked. The leaves are a glossy darker green and thick in texture. They are ovate in shape and slightly toothed along the bottom margins. They might turn yellow, orange, or red in the fall. The flowers are very small and white with 5 petals in a star shape. These flowers grow in six inch racemen.

The fruit is the same as the cherry-round, coming off of long stems, and red, dark red, or even appearing black.

Where to gather it

You might find the tree in any forest as it is fairly adaptable. In the late summer to early fall, you might be able to pluck the fruit from the tree.

How to gather it

Alternatively, you might have to wait for the fruit to fall from the tree, in which you can lay down a tarp or similar. It is best not to let them fall directly on the ground because the fruits will start rotting right away.

How to cook with it

Edible: The fruit can be used as its equivalent from the store. They can be eaten raw, put in desserts, or preserved as jams.

Wild Cherry Pie Filling

This recipe makes one pie.

Time: About 20 minutes

Ingredients:

- 5 cups pitted cherries
- 1/2 cup water
- 2 tbsps lemon juice
- ¾ cup sugar (Adjust if needed. This recipe is for wild berries which tend to be more sour)
- 4 tbsps cornstarch
- optional: ¼ tsp almond extract

Instructions:

1. In a pot, combine all ingredients **except** the almond extract.
2. Bring to a boil while stirring.
3. Cook for 10 minutes.
4. Add extract.
5. Take off heat.
6. Store in the freezer or add to pie crust right away.

Author: Just like the regular cherry, you should be cautious of the pit of the cherry because it can contain cyanide. I love to treat my wife by preparing her a chocolate cherry cheesecake with fresh cherries. She goes crazy for it!

13

BLACK LOCUST

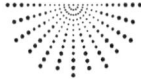

The black locust is scientifically named *Robinia pseudoacacia*.

IDENTIFICATION OF THE PLANT

The tree reaches up to 30 feet on average, but in some cases it can grow past 150 feet. It is typically very erect. The bark is very rough in texture and gray with a red undertone. The leaves are dark green compound leaves, which are 1-2 inches in length and oval shaped, noticeably more round than most leaves at the point.

The white flowers hang down in short drooping stems. The flowers are about an inch wide and are spread out across the stem alternatingly. The flowers are trumpet shaped and asymmetrical. The short individual stems are 1-2 inches and red. There is a reddish cupped bracket that holds the flowers onto the stem. There are quite a few of these drooping branches on the tree, giving it a beautiful show when it is in bloom.

Where to gather it

The tree likes lots of sun and drier areas. The bloom period is about two weeks in late spring: around May to June.

How to gather it

You can gather the flowers by snipping off the drooping flower branches. They are fairly large, so you will need to have scissors as well as a large enough basket. If the tree is not located somewhere you naturally see every day in the spring, make sure to check on the tree regularly at this time of year so that you don't miss the short bloom period.

How to cook with it

Edible: The flowers can be eaten raw, battered, or fried. All parts of the flower can be eaten, the reddish cup bracket being the sweetest part. They can be made into syrup.

FRIED BLACK LOCUST FLOWER

Time: About 30 minutes

Serving Size: 4 servings

Ingredients:

- 12 bunches of Locust flower
- 1 cup flower
- ⅔ cup water
- 1 pinch baking soda
- 1 pinch salt for taste
- 1 tbsp honey
- oil for frying (best if it's fresh)
- 1 tsp sugar

Instructions:

1. Wash the flowers in cold water and remove any inedible pieces.

2. Dry them off well.
3. Mix together flour, baking soda, salt, water, and flour to make a batter and let rest for a few minutes to a half hour.
4. Heat a pan with oil. Make sure it isn't too hot, but hot enough that it won't over cook the petals.
5. The batter should be a beautiful crunchy golden color.
6. Sprinkle it with sugar and serve.

Author: It can be a little inconvenient to time getting this plant right, but it is a beautiful and delicious plate to create. So, if you have the time and the tree, it's worthwhile to try and forage for some black locust flowers.

14
BLACK WALNUT

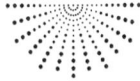

*T*he black walnut is also scientifically known as *Juglans nigra*.

IDENTIFICATION OF THE PLANT

The black walnut tree can grow up to 130 feet tall. The bark on the tree is very dark, thick, and rough textured with cracks. On younger growth, the color might appear lighter. Leaves are compound and alternating on a 1-2 foot stem with about an inch between each leaf. The leaves are more triangular in shape but elongated to four inches and narrow. Slight teeth on the margin. There are flowers in the shape of catkins in the spring but they are unremarkable.

The fruit is green to brown, round, and about three inches in diameter at their largest. Inside is the nut, which is the same as a store bought walnut nut appearance, resembling a brain with lobes and deep wrinkled texture.

Where to gather it

The black walnut tree likes low land areas with the

combination of well drained soil. The nuts are ready to harvest in the fall.

How to gather it

Unlike fruit, you can pick the nuts off the ground that have fallen from the tree. When they fall on the ground, they are fully aged. The harder part of this process is to husk the outside of the black walnut. This process may take a while and may involve heavy duty steel brushes and other tools to ensure its clean.

How to cook with it

Edible: Yes.

Once you have cleaned the husk of the nut, it needs to go through a drying process that might take a few days to completely dry. After this, the shell still needs to be broken. This can be a dangerous process if you don't have the right tools and aren't careful. Once you have broken the shell open, you need to boil all the material for about 30 minutes. This can help separate the nut and the shell, but it can also release the oil that can be used in cooking. This leaves the nut left to be used like any other nut in cooking.

BLACK WALNUT COOKIES

Time: 20 Minutes

Serving Size: 1 batch

Ingredients:

- 1 cup black walnuts
- 3 cups all purpose flour
- 1 cup white sugar, 1 cup brown sugar
- 1 cup butter
- 1 pinch each of salt and baking soda
- 2 eggs

- optional: warming spices like nutmeg and cinnamon

Instructions:

1. In a large mixing bowl, add cream, butter, and sugar, then beat in the eggs.
2. In a separate bowl, mix flour, baking soda, and salt.
3. Mix the dry ingredients into the wet ingredients.
4. Once mixed, add in the walnuts.
5. Refrigerate until firm.
6. Preheat the oven to 350 °F.
7. Roll dough into small balls and flatten. Don't play with them too much because it'll warm them up.
8. Bake for 1o minutes.

Author: The husk of the walnut is used for homemade dyes. My parents would make the 'nocino' liqueur and put it in the dark cellar to soak for months until it was done. I loved sneaking into the cellar and grabbing a little glass, filling it with this delicious alcoholic drink, adding some sugar to it, and having some sips of it. It was a great replacement for a sweet treat when my mum wouldn't get me the candy I wanted.

15

BLUEBERRY

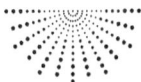

There are two kinds of blueberries that you might run into: the first being *Vaccinium corymbosum* or the swamp blueberry, and the second *Vaccinium angustifolium* or the lowbush blueberry.

IDENTIFICATION OF THE PLANT

Both species of blueberries grow in shrubs, but there is an obvious height difference, with the swamp blueberry growing up to 12 feet and the lowbush growing up to 1 foot.

On both blueberries, the leaves on the plant alternate and are simple—up to three inches on the swamp berry and half that on the lowbush berry. The small flowers cluster and hang from short stems into off-white, thick petaled bells.

The fruit on the lowbush berry tends to be slightly more blue, whereas the swamp blueberry tends to be a little darker, both with a dusty-like coat. The berries are small and round with a little jagged crown on one side.

Where to gather it

As the name swamp blueberry implies, this variation of

blueberry is more common in wetlands. Alternatively, the lowbush berry likes higher elevations and rocky, dry land. They fruit in the summer.

How to gather it

Unlike the other berries that are foraged in bramble bushes, the blueberry tends to be a **little** bit easier since it grows in a more manageable shrub. You can pluck the fruit right off of the plant when it is ready and taste test to see if they are right.

How to cook with it

Edible: Yes, the fruit.

Blueberries can be treated like any blueberry you find in the store. This means they can be eaten raw, preserved, dried, or cooked into dishes of your preference.

BLUEBERRY MUFFINS

This recipe makes 10 regular sized muffins.

Time: 30 minutes

Ingredients:

- 1 ½ cups all purpose flour
- 3/4 cup white sugar
- 2 tsps baking powder
- 1/3 cup oil (canola or vegetable)
- 1 large egg
- 1/2 cup milk of choice
- 1 ½ tsp vanilla extract
- 1 cup fresh or frozen blueberries
- a pinch of salt

Instructions:

1. Preheat the oven to 400 °F.

2. Butter the muffin tin or use cupcake liners.
3. In a bowl, mix together liquids. In another, combine the dry ingredients.
4. Combine the two, mix only as much as necessary. The trick to fluffy muffins is to avoid over mixing. Add more milk if necessary. The batter should be fairly liquidy, unlike cookie recipes.
5. Add in blueberries, then fold the batter over lightly.
6. Pour or scoop batter into muffin tin.
7. Bake for 15 minutes, do a toothpick test. Then bake for another 5 minutes and try again.

Author: While blueberries are obviously blueberries when you are careful, always take some extra caution when it comes to eating strange berries. A lot of wild berries are the same shape and, without care, might be misidentified.

Also, wild blueberries aren't comparable to the ones you find in the supermarket. Make sure to sit down next to blueberry bushes and enjoy the sweetness of these delicious berries.

16

BRAMBLES

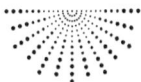

RASPBERRIES, BLACKBERRIES, AND BLACKRASPBERRIES

*B*rambles are an umbrella term for plants that are able to be more shrub-like in some environments and, in others, might climb like a vine or just be a thick brush. It is easier to group them together in these cases because the plants themselves look similar, and it's a matter of identifying which berry you have in front of you. In this chapter, when we are talking about brambles, it's specific to the rubus genus.

IDENTIFICATION OF THE PLANT

Brambles: The shape of each of these plants are brambles: long thin stems that are semi-erect, sometimes creating an arch shape once the vine becomes long enough and droops.

Raspberries are a small red berry that looks like a hollow stack of small balls. It has little hairs that come out of the fruit. Similarly black raspberries also share a resemblance to raspberries but are different in color.

Blackberries are solid, not hollow, and have a glossy look to them. Black berries, while similar to raspberries, are also slightly bigger and sometimes longer than a raspberry. While they are both edible, it's important to know the distinction. When blackberries are not ripe, they are red and can look like a raspberry, but it is not recommended to eat blackberries until they reach maturity.

Where to gather it

Brambles like a lot of sunlight and some shelter from other harsh elements. This means you might find brambles in openings or on the edge of woods. Keep in mind they also do not like to sit in any water, so look for higher grounds.

How to gather it

Pick as fruit ripens. It is important to take some caution when dealing with vines or shrubs because they can be prickly. Since bramble plants do climb and spread, you might not be able to reach the entire plant for harvesting. If you are growing the plants yourself, you might have more luck training the plant for ease of access.

How to cook with it

Edible: The variety of berries that come from brambles is the same as the berries that you get in the store. They can be eaten raw or used in a number of dishes, mostly desserts. They can also be preserved in jam.

Mix Berry Jam
   Time: 40 minutes
   Serving Size: 6-10 jars
   Ingredients:

   - 6 cups of fresh berries of your choice

- 5 cups white sugar (adjust to your sweetness preference, but remember that sugar is what helps preserves, so don't undercut too much)
- 1 tbsp lemon juice

Instructions:

1. Put a small plate in the freezer.
2. In a large pot, add the berries and the lemon juice.
3. On medium heat the berries will slowly come to a boil and then soften.
4. Bring to a boil and reduce heat to a simmer for 15 minutes.
5. Add sugar off on low heat and stir until dissolved. Then bring back to a boil.
6. Boil for about 5 minutes.
7. Take out the plate in the freezer, put a small drop of jam to cool it quickly and test for jam consistency.
8. Once done, skim the top with a ladle for impurities and foam.
9. Add to sterilize jam jars.
10. Seal the jars and label.

Note: Once the jars are sealed, take the ring off of the jar.

Author: While thickets that are full of brambles can be less than fun to walk through, it is a great natural wall in your own gardens as well.

At my grandparent's table in the summer, they'd always have a bowl filled with these delicious berries in their fridge. I was always excited to drive to my grandparents, open the fridge, and put them on top of my favorite banana yogurt.

17

BURDOCK AND RHUBARB

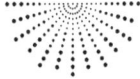

urdock is the common name for the genus *Arctium*. Both common burdock, scientific name *A. minus*, and greater burdock, scientific name *A. lappa*, are the most used when it comes to eating. Both are common in North America.

While Burdock and rhubarb are not part of the same family, rhubarb belongs to the *Rheum* genus, it makes sense to put them together because of how similar they look. Young burdock and rhubarb are almost identical, especially to those who do not know the difference, and the mistake is not one that can be overlooked. Rhubarb leaves are not edible and can make you pretty sick. Also, if you are trying to forage rhubarb stalks for a pie and end up with the celery-like burdock, you will find yourself with a less than pleasant dessert.

IDENTIFICATION OF THE PLANT

A. minus can grow up to 6 feet and *A. lappa* to 9 feet.

Very large leaves, spanning up to 20 inches long and 12

inches wide, in a heart shape. Its leaves have a wavy edge with thick, fleshy, and veined stems. The leaves are slightly textured by veining. *A. lappa* has solid stems, *A. minus* has hollow stems. When mature, they get a woody stem that gives them height and may have a reddish tone to it.

The difference in rhubarb and burdock at this stage is subtle. The leaves on a rhubarb curl or wrinkle up on the edges, while the Burdock's leaves have a slight hair on the underside. The stalks of a rhubarb are solid stems. Rhubarb's stems are also noticeably red or pink—but this should not be your only indication, as burdock can have a (typically not as bright) red hue.

These plants are known for their burrs, which contain their tiny purple/pink flowers at the top with long, white stamen. The burrs are green and brown when dried. They are covered in what looks like one inch thorns but have a small hook on the end for clinging on to a passerby.

One of the easiest ways to tell the difference between the two is that *A. minus*'s burs are usually singular, whereas *A. lappa* grows them in clustered florets. The other differentiator is that *A. lappa*'s leaves will be rounder, whereas *A. minus* will have slightly elongated leaves.

Where to gather it

Burdock is typically considered a weed, and finds itself in common weedy areas such as disturbed, grown over places. Best to harvest in the spring because the plant can get woody. Rhubarb is typically harvested mid spring and not preferred past July. Rhubarb likes full sun and well draining, fertile soil.

How to gather it

The young burdock plant can harvest the roots in the first year as well as the base of the leaf stems. In the second year, you can harvest the flower stalk. Use a sharp pair of scissors to sever thick vessels. Rhubarb can be broken off or cut at the base, removing the leaf. The stalk does not get completely red, so do not use that as an indicator. They are the best when they are about .5 to 1 foot long.

Dig the roots in the summer. They can be deep and hard to get out, so be prepared with the right equipment if you are planning to harvest.

How to cook with it

Edible: Yes, Burdock roots, leaves, and stalks. Rhubarb stalks, but NOT the leaves.

Burdock: For the roots, peel and slice like a potato. Boil for 20 minutes, change water and boil until tender. The flavor is described as crunchy, earthy, and nutty. Flower and leaf stalks similar to celery should be peeled, then used raw or boiled until tender.

Rhubarb: The stalks can be frozen. They are mostly used in pies or made into syrups.

Roasted Burdock Root
Time: 30 minutes
Serving Size: 4 servings
Ingredients:

- 1 pound Burdock root
- 3 tbsps Olive Oil
- 1 pinch salt, pepper, and other spices for taste
- optional: foraged field garlic, garlic mustard, wild leeks, or wild onion for taste.

Instructions:

1. Preheat the oven to 400 °F.
2. Prepare the root by washing, peeling, and cutting it up into slices about 1 inch wide.

3. Spread the root out on a tray and evenly coat them with oil and spices.
4. Cook for 10-15 minutes, flip, and cook for another 10 minutes.
5. Recommended: a splash of soy sauce.

Author: Burdock has a long history with herbal medicine, mostly being used as a blood purifier, among other things. The root is high in antioxidants. The burs of the burdock evolved to catch on passerby to spread their seed farther.

This makes the most delicious addition to a pumpkin soup. I like to sprinkle the roasted roots on top to bring out a very delicious zesty flavor! Just had this for lunch exactly as it's written in the recipe and you can't beat it.

18

BUTTERNUT

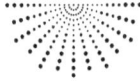

The scientific name for the butternut tree or white walnut is *Juglans cinerea*. It is illegal to gather or have in your possession in the state of Minnisota, but legal in every other state.

IDENTIFICATION OF THE PLANT

The butternut tree can grow up to 120 feet but mostly keeps to half that height. The bark is light gray, very deeply grooved, and thick. The leaves are alternate and pinnate, the main stalk being about 2 feet, and the leaves being up to 4 inches long with 19-21 leaves on the stalk. It will have a leaf at the end of the leaflet, meaning there should be an odd number of leaves. The shape of them is very simple, with a main vein down the middle and straight veins coming out of it parallel. The flowers of male and female trees are both rather unremarkable.

The fruit is oval and smaller than a lime, with green skin around the nut.

Where to gather it

The tree likes lowland forests. You can forage immature nuts in the summer for pickling or mature nuts in the fall when they fall from the trees.

How to gather it

If you are harvesting the young seed, pluck it off the tree in the summer. If you are going for the mature nut, pick them off the ground. They need to go through the same process as the walnut, being cleaned of the outer husk, dried completely, and then, after about a week, need to be removed from the shell before cooking.

How to cook with it

Edible: Yes, the nut.

Once you crack the nut, boil all parts, sift the oil off the top, and remove the nut from the shell. Use the oil for cooking and the nut for whatever nut recipe you choose. The younger nuts picked in the summer can be pickled after they are cleaned from the fuzz and boiled until the water runs clean.

White Walnut and Cleavers Pesto

This recipe makes 1 ½ cups of pesto.

Time: 20 minutes

Ingredients:

- ½ cup white walnuts/butternuts
- 2 cups cleavers
- ¾ cup parmesan
- ½ cup olive oil
- 2 cloves garlic, adjust for your preference
- pinch of salt and pepper
- 4 tbsps lemon juice

Instructions:

1. Wash cleavers and dry.
2. Add everything into a blender or food processor.
3. Blend until it's a thick liquid, but it doesn't need to be completely smooth.
4. Taste and adjust for your preference if needed.
5. Add to food like pasta or sandwiches.

Author: It is a good idea to expand your foraging from fruits and greens to nuts because they are able to have a longer shelf life and add a more diverse nutritional aspect to your plate. In foraging, it can be hard to find good sources of protein. While you should eat everything in moderation, if you are trying to supplement more of your diet from found food, nuts are a really good choice.

19

CATTAIL

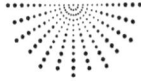

attails or bulrushes are a common name for the genus *Typha*. The common foraging types of cattails are *T. latifolia* and *T. angustifolia*. Like the Amaranth, it filters the ground and water, so be cautious of where you are harvesting from.

IDENTIFICATION OF THE PLANT

Cattails are fairly easy to identify because they have a fairly unique look and their swampy home. Standing up to nine feet in some cases, cattails have their iconic brown catkins. The catkins are up to 1.5 feet long and fairly narrow at the top of a long stem, often being referred to as corn dogs or cigars. The catkins are soft and velvety on the outside, and when broken open is a very densely packed cottony seed that can burst with pressure. The catkin is the flower of this plant. The catkins on *T. latifolia* are longer and wider than the *T. angustifolia*.

The leaves alternate and are very long, growing up to eight feet, but are also narrow and only about one inch wide.

Another way to tell the difference between the two species is that *T. angustifolia*, the narrow leaf cattail, has narrower leaves. The leaves are sheathing at the base. The leaves have parallel veining that give the thick leaves a coarse texture. They have smooth edges and come to a point.

Be cautious of water irises, as they look very similar to the cattail in foliage. To identify, you can notice the catkin on a plant that has matured. If you are identifying at a different time in the year, the cattails will have a round stem with the leaves wrapping around, versus the water iris, which is flat and fanning.

Where to gather it

Typically found in shallow water like ditches and ponds. You can harvest the root any time of the year but you can get the best results in fall and winter. Shoots and catkins for eating have to be harvested in spring before they mature. The flowers will bloom between June-August. Starting in July, the pollen can be collected after flowering.

How to gather it

Make sure to have waterproof boots and gloves. If there are little to no shoots to pull the ribosome out, the roots can be harvested by digging. The ground might be soft due to the swampy conditions that they grow in, but you should still bring a shovel along. Once the ground is loosened, simply pull up.

When gathering the shoots in the spring, you can snap or cut them off from the ground or the root. It is better to get the younger shoots because the older the plant is, the less preferable edible shoots there are. You can cut off the leaves and roots if you are not using them.

The catkins can be harvested, but timing it is difficult. In the spring, you can find green female catkins hidden amongst the leaves.

The pollen can be harvested by shaking the flower into a

bag or other catcher. This might be done by keeping the plant intact, or carefully cutting it (so as not to shake out the pollen), and shaking it into a catcher to have more control over the movement.

How to cook with it

Edible: Yes.

The root needs to be cleaned, although only the main ribosome should be used and not the branching root growth, but you can use the sprouts of greenery. You can cook the root by boiling or roasting it. The root can be turned into a flower by being dried, cleaned, and grounded into powder.

The green catkins can be eaten like corn on the cob. The pollen can be used as flour. Young shoots can be eaten like asparagus. Unwrap the thicker pieces along the outside and cut off where the leaves turn green and flat. What is left is about a foot of narrow, white stalk. They can be eaten raw but are preferred cooked and roasted.

FRIED CATTAIL SHOOTS

Time: 10 minutes
Serving Size: 2 servings
Ingredients:

- 1 cup Cattail shoots
- 5 tbsps vegetable oil
- garlic (adjust to preference)
- 1 pinch salt and pepper for taste
- 1 pinch sesame seeds

Instructions:

1. In a pan, warm up the oil and spices.
2. Cut and wash the cattail shoots.

3. Turn the heat up to high and add the shoots.
4. Fry for 5 minutes, stirring occasionally.
5. Add this to rice or other stir fry mixes.

Author: Cattails are a great plant for the forager to remember because the whole plant can be useful; it's multi-purposeful and is beneficial for different qualities all year long. If you are a forager for other purposes besides food, the cottony seeds are useful for stuffing and, when cleaned, can be used for a cotton ball replacement. The leaves have also been used in weaving. The gel found between leaves is used similar to aloe vera.

20

CHICKWEED

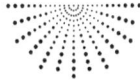

*C*ommon chickweed, also known as Stellaria media, is a great plant to have in your foraging arsenal because it is available all year long. *S. jamesiana* (James chickweed), common in the west, *S. holostera* (Greater stitchwort), and *S. graminea* (Lesser Stitchwort) are also part of the *Stellaria* genus and can be used in foraging. Also mistaken for the chickweed is the *Cerastium* genus because they have much more noticeable hair along the stem. Some warn against *Euphorbia maculata* as well, which will ooze a white cap when cut at the stem.

IDENTIFICATION OF THE PLANT

The common chickweed is more of a ground cover and doesn't get more than a few inches off of the ground. All four species can be identifiable by their white flowers that are a little less than half an inch wide. There are five petals that do not touch, but are divided about halfway down, giving it the look of two petals fused together or even the look of two completely different petals.

The leaves on the common chickweed are green, can be more dense, and might be almost succulent-like. They are smooth and oval and sometimes look more round with a slight wave. The closer to the ground, the longer the leaf stalk; if you look at the top of the plant, there is almost none. They come out of the stem in an opposite pattern. The bracts of the flower have noticeable hairs that can be slightly seen on the leaves and a single row down the stem. This is also noticeable on unopened flowers. The stem can be green or have a tinge of red.

Both Stitchworts reach about 1 foot, have a very delicate, hairless stem, and branches angularly. The leaves are opposite, about an inch long, narrow, triangular, and smooth edged.

Where to gather it

Common chickweed is considered a weed and is more likely to pop up in places that have been disturbed and well watered. Common chickweed is available all year. The stichworts are available in spring for the greater stitchwort, and spring through summer for the lesser stitchwort. Stitchworts prefer more medowie areas or woodlands. Since they winter well, by the spring they are already established.

How to gather it

For all species of this plant, cut off the younger, non flowering parts of the plant with scissors. Look for the tender leaves. It's best to harvest right after it rains to get a fresh and perky harvest. If you are looking for it in the winter, it might be hard if your area has a lot of snow. It helps that it likes shaded areas, which might mean less snow cover. If you know where you can find some, you might be able to harvest these greens all year.

How to cook with it

Edible: Yes.

The stitchworts and chickweed are both used as greeny in

salad, or on its own. The leaves and the flowers are used, but you can also boil the green shoots. Boil for a few minutes as it is delicate.

SAUTEED CHICKWEED

Sauteed Chickweed can be eaten as a side or as a topper for rice or chicken.

Time: 10 minutes

Ingredients:

- 2-3 cups Chickweed
- 1-2 tbsps butter (adjust as necessary)
- add spices like salt, pepper, garlic, onion, etc.

Instructions:

1. In a pan on medium high heat, soften onion and garlic.
2. Add any other spices you like.
3. Wash chickweed and remove wilted pieces.
4. Add chickweed in a pan for about 1 minute and not much longer.

Author: Chickweed is used for treating itchy skin but also contains saponin, which can cause some people in excess doses to be sick. It's a favorite amongst foragers, not only because it has a great nutritional value, but also because it has a great taste and is similar to spinach.

21

CHICORY

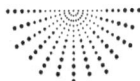

*C*hicory, also known as *Cichorium intybus,* is a well known coffee substitute. As a forager looking to source as much as you can locally and naturally, this plant might be your solution to having your morning pick-me up.

IDENTIFICATION OF THE PLANT

Chicory is most identifiable by its blue, pink, or white flowers in the summer. The flowers are about 2 inches wide. The flower is flat like a daisy, with petals that come to a flat end with a jagged edge. The plant itself can grow up to five feet tall, but its stem is sparsely covered with leaves and is thin and spindly. It might have a few branches. The leaves at the base are very similar to a dandelion, with a long stalk tinged with red jagged edges, and it starts narrow and gets wider towards the end of the leaf with a roundish tip, reaching up to 12 inches. The leaves up the stem alternate and are significantly smaller (maybe four inches long), narrow triangle shaped, and smooth edged. Chicory has a milky sap that can be seen when the plant is damaged.

Where to gather it

Chicory is a weed that likes disturbed areas, meaning it is commonly found on the side of the road. It's best not to harvest from near roads but think of similar areas. Gather roots fall through spring, and harvest leaves in early spring.

How to gather it

In fall through the spring, you can harvest the roots. Locate the plant and bring a shovel. You will have to dig to get to the root. For leaves, the earlier in the season the better, since older leaves become more bitter. You can simply pluck the leaves off the plant. The leaves at the base of the plant that look like dandelions are the preferable leaves.

How to cook with it

Edible: Yes.

Boil or chop and eat raw crown base leaves, comparable to other leafy greens. For the root, clean, roast, and grind up like coffee grounds.

CHICORY COFFEE

Time: up to an hour

Serving Size: 1 cup

Ingredients:

- 1 cup chicory
- 1 cup water
- optional: milk, sweetener, and cinnamon for taste
- optional: coffee grounds

Instructions:

1. Wash your chicory root well.

2. Cut your chicory root into 1 inch pieces. Optional: you can mince them to make the grinding part later easier.

3. Heat the oven to 350 °F.

4. Lay out the roots evenly on a tray.

5. Put in the oven to roast. The timing depends on the thickness of your chicory pieces and your oven. It will be approximately an hour and a half. A good indicator is when you are able to smell the aroma.

6. Take the roasted roots out and let them cool.

7. Grind the roots. Like coffee, the finer the grounds, the stronger the coffee. Don't ground too much, it should not be powder thin.

8. If you want, you can use this on its own or mix it with 1:1 with your usual coffee grounds.

9. Add to a coffee maker, french press, or however you make coffee.

10. Add milk and sweetener to your preference.

Author: Chicory also has a reputation for treating worms/parasites.

22

CLEAVERS

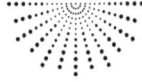

The scientific name for cleavers is *Galium aparine*.

IDENTIFICATION OF THE PLANT

This plant is herbaceous and only semi-erect. The leaves come out in 8 out of the plant in rows fanning straight out. There will then be an empty stalk for a few inches, and then another fan of leaves. These leaves are narrow with slight fuzz. The flowers bloom in threes and are small and white. The fruits are two ¾ circles fused together with hooked hair.

Where to gather it

The best place to find clevers is in moist areas, like lakes and fields with rich soil. While I recommend harvesting this plant in the spring, you can also harvest the fruits in the summer.

How to gather it

It is best to pick the stalks when the plant is younger and tender. Once identified, you can cut and collect what you need.

How to cook with it

Edible: Yes, the young stalks can be steamed or boiled. The fruits can be dried, ground, and used as a coffee bean substitute.

AUTHOR: CLEAVERS HAVE A LOT OF LOOK-ALIKES, INCLUDING goosegrass and more, but they all fall into this same category.

23

COMMON ELDERBERRY

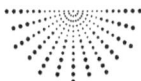

The common elderberry is scientifically known as *Sambucus canadensis*.

IDENTIFICATION OF THE PLANT

The common elderberry is a shrub that grows to about nine feet tall. The leaves grow in an opposite pattern with up to nine leaves on leaflets. The leaves are long and narrow to a point, standing at about four inches long. They can look like they have been folded down the middle.

Its flower is a round, flat top floret with small white flowers, about up to 10 inches across. The individual flowers are star shaped with noticeable stigma. The fruit is a round berry but is smaller than a blueberry. It is round, smooth, and black.

Where to gather it

It likes sunny locations and is fairly adaptable. The berries are ripe in the late summer.

How to gather it

When the berries are fully dark purple or black, you can pick the berries off of the bush.

How to cook with it

Edible: Yes, the berries can be eaten. However, they should not be eaten raw because they are poisonous. The rest of the plant is also poisonous.

ELDERBERRY SYRUP

Time: 40 minutes

Ingredients:

- 2 cups water
- 1/2 cup elderberries
- ½ cup honey
- optional: spices, ginger, cinnamon, and cloves

Instructions:

1. Clean the berries.
2. In a pot, add everything but the honey.
3. Bring everything in the pot to a boil.
4. Once it hits a boil, bring to a simmer for 20-30 minutes. This should reduce the water by about half.
5. Take off heat and strain out any elderberry flesh and skin.
6. Let cool and add honey.
7. Store for 2 weeks in the fridge.
8. The consistency is not thick, but adding sugar instead of honey as it boils would create a more syrupy texture. This also extends its shelf life.

Author: Elderberries are packed with antioxidants and can boost your immune system. Even though it's a lot of work to prepare the berry, its ability to heal your body against colds is worth it.

24

COMMON MALLOW

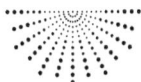

*ommon mallow, also known as dwarf mallow, is scientifically called *Malva neglecta*. There are other plants in this genus that have more decorative flowers that you might see more often in gardens. Some species are used like lettuce. Europeans call *M. Sylvestris* common mallow, but this shouldn't be an issue foraging in the North East, as it is local to the middle east.

IDENTIFICATION OF THE PLANT

Common mallow is more weedy looking than the others of its genus that have more flashy foliage and flowers. This plant stays closer to the ground but can grow up to two feet. The flowers have five petals and can come in a variety of colors. Some flowers of this genus take on a more star shaped look, with more narrow petals, while some have bowel shapes with round, overlapping petals. The common mallow has a star-shaped flower that is white and can be tinged with pink.

The leaves alternate and are palmately lobed. They are

rounded with 7 lobes and are slightly ruffled in texture, with jagged edges. The leaves come off longer herbaceous stalks.

Where to gather it

Common mallow is a common weed that likes areas with more moisture and full sun. It can be found in the summer, fall, and winter.

How to gather it

When you have identified the plant, you can clip off the leaves, or you can dig up the root and harvest the whole plant. It flowers in the summer, but roots later in the season.

How to cook with it

Edible: All of this plant is used. The seeds can be raw or cooked like rice. The leaves and flowers can be used in salads, but it's best not to eat it alone. When cooked, they turn into a mucus-like texture that is used for thickening. The root is boiled to create a mucus as well as a meringue, egg white substitute.

MALLOW MERINGUE

This can be used on pie, meringue cookies, etc.

Time: About 30 minutes

Ingredients:

- ⅓ cup Mallow root or seeds
- 1 cup water
- ¼ tsp cream of tartar
- 1 egg white
- ¾ cup sugar
- 1 tsp vanilla extract

Instructions:

1. Wash and chop mallow root or peel the seeds.

2. Add dried root to water in a pot and bring to a boil.
3. Reduce water by half (you may or may not see the thickening).
4. Remove from heat and strain.
5. In a mixing bowl, beat the eggwhite until white.
6. Add cream of tartar, take ½ cup mallow thickener slowly add to mixing bowl, beating the mixture as you go.
7. Once mallow is all mixed in, repeat the process by slowly adding in the sugar and finally the extract (optional).
8. Beat until foam is fairly firm and fluffy.

Author: The plant is undesirable to livestock owners as it is toxic to graziers. The leaves have been used for sore throats and digestion.

COMMON MILKWEED

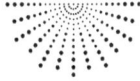

*M*ilkweed is also known as the butterfly or silk flower, but it's scientifically known as *Asclepias syriaca*. The reason it is called the butterfly flower is because of the Monarch butterfly. The larvae or the butterfly *only* eat milkweed. Because of this and their decreasing population, be conscious of how much milkweed you take and whether it is necessary to take that much. Other species of milkweed and young dogweed can look like this plant. Dogbane has branches and no hairs, and other milkweeds have a more narrow leaf.

IDENTIFICATION OF THE PLANT

The plant grows up to six feet tall and mostly erect. The leaves are simple, smooth, and elongated. They grow up to 11 inches, have a prominent main vein, and a soft almost hairy underneath. The leaves are thicker, allowing them to be more erect. When damaged, the plant has a latex white sap.

The flowers are usually pink in florret clusters that are round, 1-5 clusters on each plant. The flowers are on short stems that originated from one center. The individual flowers are star shaped with prominent staples that open wide below the tiger shaped flower.

Where to gather it

The plant likes well dried soils with lots of sun. This plant is found in places that look more weedy, like ditches and meadows. The plant can only be eaten when it's young, so it's harvested in the spring.

How to gather it

Harvest shoots by cutting the young, tender plant before it matures.

How to cook with it

Edible: Yes, the young shoots, leaves, buds, flowers, and fruit are edible. The mature plant and fruit is toxic.

MILKWEED SHOOTS

Time: 25 minutes
Serving Size: 4 servings
Prep Time: 5 minutes
Cook Time: 20 minutes
Ingredients:

- 15 young shoots
- 2 cups water
- 2 tsps butter
- pinch of salt and other spices of your preference

Instructions:

1. Bring a large pot of water and salt to a boil.
2. Add shoots.
3. Boil for about 10 minutes, don't recuse water.
4. In a pan, warm up butter, garlic, onion, etc.

5. Add boiled shoots to a pan for a few minutes until you see a nice crisp forming.

Author: Milkweed is a favorite among many bugs because it's a good pollinator. Its flowers are very sweet and that is why it's their favorite snack.

26

COMMON STRAWBERRY

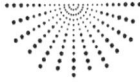

*lso known for its latin name *Fragaria virginiana*, strawberries are a great example of how fruits can exist in the wild without being genetically bred. They do have a look-alike that might be called the mock strawberry or Potentilla indica. The plants are very similar, and although the claims that they are poisonous are wrong, they are still not edible, in the sense that it is hard and tastes undesirable. The flower is yellow, the fruit is more round, and the seeds stick out like spikes instead of fitting nicely in the side like the wild strawberry.

IDENTIFICATION OF THE PLANT

Of course the most identifiable part of the strawberry plant is the fruit. Like the strawberry that you find in the store, it is red. In the wild, however, they are much smaller, about the size of a raspberry. The flowers, at about .5 inches wide, are white and star-shaped, with five petals and a yellow round center. The plant is more of a ground crawler and doesn't have a main stem, but it does reach out with tentacles that spread the plant. The leaves cluster in threes, are oval shaped with very jagged edges and a deep folding texture with a slight fuzz underneath.

Where and how to gather it

The strawberry can be found in the woods and in more open areas too like meadows. They are available in the spring and well into the summer.

They are small, so keep your eye on the ground. The red fruit can be easier to spot. The fruit can be plucked right from the stem, and the leaves can also be harvested any time of year.

How to cook with it

Edible: Strawberries can be used in the same ways that store bought strawberries can be used, meaning that they are very versatile in use: pies, jams, or eaten raw. The leaves can be dried and used for tea.

STRAWBERRY AND RHUBARB PIE

Time: 45 minutes

Serving Size: 8 servings

Prep Time: 5 minutes

Cook Time: 40 minutes

Ingredients:

- 3 ½ cups chopped rhubarb
- 2 cups strawberries
- 1 cup sugar
- ½ cup flour
- 1 pie crust
- 1 tbsp butter

Instructions:

1. Preheat the oven to 400 °F.
2. In a bowl, mix flour and sugar.
3. Add cleaned and chopped rhubarb and strawberries.
4. Add mixture into pie crust, cover top with crust.
5. Add holes to the crust top, add little pieces of butter, and sprinkle with sugar.
6. Cook for 40 minutes.

Author: Strawberries can be a really great food to forage for because they can also be preserved fairly well in jams. While fresh strawberries can be enjoyed in the spring and summer, they can also be enjoyed all year round. They are high in vitamin C and a great pick-me-up in general.

27

CORNELIAN CHERRY DOGWOOD

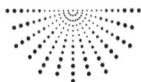

*C*ornelian Cherry Dogwood is known scientifically as *Cornus mas*. The fruit is similar to cranberries and cherries. It is considered invasive. This tree looks similar to Forsythia; it has similar yellow flowers but does not grow bright red fruit.

IDENTIFICATION OF THE PLANT

The small tree or shrub is actually quite attractive looking. The silhouette of the tree is similar to that of a large bonsai tree. The branches at the top spread out wide, but aren't as tall. The older growth of the bark is a dark brown, while the new growth is a green. The leaves are long ovals that reach up to four inches long at a point. They are slightly glossy and curl in towards the center from the sides.

The tree puts on a show in the late winter/early spring when it blooms. It becomes covered in clusters of little yellow flowers. There are no leaves at this point so it's just the yellow flowers. The individual flowers have 4 petals each and cluster in round florets on opposite branches.

The fruit is oval and olive-like with red or yellow berries. The fruit hangs in very short stems in clusters of 1-5.

Where to gather it

The plant can be found in woodland areas and public areas with dry soil because it's ornamental. The fruit is ready in the late summer.

How to gather it

Harvest the berries when they are bright red by plucking them off the tree.

How to cook with it

Edible: Yes, the cooked fruit is edible.

Mostly made into jams and mizde with other sour fruit; it can also be dried.

CORNELIAN CHERRY MARMALADE

Time: 22 minutes

Prep Time: 2 minutes

Cook Time: 20 minutes

Ingredients:

- 2 cups Cornelian cherries
- 2 cups water
- 1 cup sugar
- a pinch of ground cardamom
- 2 tsps lemon
- 1 tbsp orange zest

Instructions:

1. Bring water and fruit to a boil.
2. Cook for 10 minutes.
3. Strain with a large hole strainer and squeeze as much juice and flesh as possible out of pits and

other impurities. Unlike jam, juice, and jelly, marmalade wants some of the flesh.

4. Add sugar, lemon, cardamom, and orange to the juice in a clean pot.
5. Bring to a light boil, reduce heat slightly and keep a steady light boil for about 10 minutes.
6. Add marmalade to jars, seal lids, and label.

Author: I always recommend taking note of plants that are attractive and grown for ornamental purposes. People are more likely to plant these attractive plants in public places like roads, parks, parking lots, etc. These are plants you can forage for in more accessible areas without needing to head into the forest.

28

COW PARSNIP

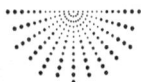

ow Parsnip is known as *Heracleum maximum*. **Be careful** when handling this herb. The sap can react to the skin and cause a rash. For some the rash can be severe. Use gloves and tools when handing. This plant can be confused for giant hogweed, which is scientifically known as *Heracleum mantegazzianum*. Luckily hogweed is larger and the stem has a purple hue. If you have severe reactions, avoid consuming.

IDENTIFICATION OF THE PLANT

The plant grows up to seven feet tall and looks more like a small bush, almost comical. The stem is hollow and covered in hairs. The leaves are very big, being up to a foot and a half long. The leaves are deeply lobed and jagged on the edge.

The flowers branch off in very large florets that are round and flat. These large florets are made up of branches that jut straight out of a centralized spot. They are composed of individual white flowers. The florets can be eight inches wide. The white flowers have four deeply lobed petals.

The seeds are half an inch long.

Where to gather it

Harvest the plant when it is young. Young shoots in the early spring. Mid-spring harvest stalks and leaves. It thrives in high moisture areas with shade, but it is also adaptable. It likes well drained soil like sand, and blooms late spring.

How to gather it

The plant can be harvested by choosing the youngest plants that are tender and green. The flower and blossoms can be cut off the plant but avoid touching the sap from the stem. The seeds can be harvested by gently shaking the seeded plant.

How to cook with it

Edible: Yes, young plants, seeds, and flowers.

Shoots, stalks and leaves can be cooked like other leafy greens or celery. Flowers and flower blossoms can be fried or eaten raw. The seeds can be harvested, frozen, and even used as spice.

The plant is best used as a herb.

Fried Cow Parsnip Buds

Time: 15 minutes

Serving Size: 4 servings

Prep Time: 10 minutes

Cook Time: 5 minutes

Ingredients:

- 1-3 cups Cow parsnip buds
- 1 cup flour
- 1 egg
- 1 pinch salt and pepper for taste
- ¾ carbonated water
- 1 cup vegetable oil

Instructions:

1. Wash the parsnip buds thoroughly, then dry them.
2. In a bowl, combine wet ingredients and in another the dry ingredients.
3. Heat the pan with oil on medium high heat.
4. Coat the buds in the wet mixture, then dry. For extra crisp, wet the dried coating and put it back into the dry coating again.
5. Add buds to the pan, but don't overcrowd the pan.
6. Cook until gold and crispy.
7. Serve with soy sauce or the like.

Author: Plants like cow parsnip have a lot of look-alikes. The flower that comes from the plant, although it has a lot of its own distinctions, can be easily misidentified. It's important to be aware of the fact that this type of flower is common and can be found with some friendly and some not so friendly plants.

CRANBERRY

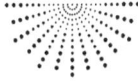

*T*he cranberry is a genus of fruit bearing plants, the name of this genus is called *Vaccinium*. There are multiple cranberry varieties that are used to harvest the edible fruit. The cranberry is related to blueberries and huckleberries. The highbush cranberry is actually not part of the same genus, being scientifically named *Viburnum trilobum*. See the highbush cranberry on chapter 56.

IDENTIFICATION OF THE PLANT

The cranberry plant can vary a bit, ranging from a shrub to a ground creeping vine. The branches then to be thin in these plants with leaves that are simple that alternate. The plant is evergreen with dark glossy leaves. The leaves are thicker. Some of the leaves are rounded oval, or more elongated.

The flowers on *Vaccinium macrocarpon* have a long bracket that sticks out from the back of the flower, as it drops from pink stems. They are a darker pink or red. The flowers have 4 petals white/pink that curl back to the bracket A long dark center spike protrudes out an inch. The flowers on *V. vitis-idaea* are small and bell shaped.

The fruit is the common denominator, round red berries about .4 to 1 inch in diameter.

Where to gather it

The plant can be found in and near swamps, lakes, flood lands and the like. They are ready in the late summer and into the winter. They like acid soils and water.

How to gather it

The plant is scooped with a cranberry scoop. The bush is typically ground cover. If you pick an accessible place to forage them, you can just kneel down and pick the cranberries off the bush.

How to cook with it

Edible: Yes, the ripe berries can be eaten.

Cranberries are commonly consumed in cranberry sauce or juice.

CRANBERRY JUICE

Time: 20 minutes

Serving Size: 9 servings

Prep Time: 5 minutes

Cook Time: 15 minutes

Ingredients:

- 1 cup sugar
- 1 cup water
- 4 cups cranberries

Instructions:

1. Wash cranberries well.
2. In a large pot, boil water and sugar.

3. Boil until sugar is dissolved.
4. Add cranberries.
5. Bring to boil, then turn down to simmer.
6. When the cranberries are done, they will pop open.
7. Add any extras for your preference.
8. Add to jars, seal, label, and let cool.

Author: If you are a new forager, making cranberry jam for Thanksgiving or other holidays in the fall can be a great project to take on. The recipe is fairly easy since cranberries are in season, and it can be a special touch for the meal, without too much stress.

30

CURLY DOCK

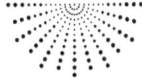

*C*urly dock is also called curled or yellow dock. Its scientific name is *Rumex crispus*. Others in this family are edible, but curly dock is the most preferable.

IDENTIFICATION OF THE PLANT

Curly dock looks weedy and grows up to five feet tall. The leaves grow from a center point in the ground like dandelions. The leaves are much longer than dandelions, as they grow up to 10 inches, and are not lobed or scalloped. The leaves have a wave to them, and the edges look scalloped but are just curling back and forth, hence the name. The leaves have a prominent main margin.

The flower comes up in separate stems. They create a flower spike in a densely packed, green to burnt orange color. Branching into long, narrow smaller florets. The individual flower is rather indistinguishable.

The fruit is .2 inches wide; it is the burnt color that encloses three seeds.

Where to gather it

The plant is a common weed, so it grows in roadsides and overgrown areas. Leaves can be harvested from spring to winter. Seeds can be harvested in the fall.

How to gather it

Harvest the best looking leaves, preferably the youngest, whenever needed. In winter, they might be yellowed but are still fine. The seeds can be cut or cleaned off of the plant when they are ready.

How to cook with it

Edible: The leaves are cooked, similar to beets. The older the leaves, the more cooking it needs. Seeds are edible when made into a flour, but the effort is typically not worth it.

Use in other leafy green recipes.

Author: Yellowdock can be difficult to identify at the beginning of the spring, and since there are so many leafy greens that are available at this time, I like to stick to yellow dock in the winter, especially since it can be so much harder to access leafy food during this season.

31

CURRANT

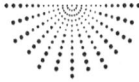

*C*urrant is a common name for a genus of plants that are scientifically called *Ribes*. A common look-alike is the gooseberry. An easy way to tell the difference between the two is that the gooseberry has thorns. It was illegal to grow currants in the US because they carry diseases that can be a danger to other plants like the pine.

The blackcurrant *R. nigrum* is a popular currant in the Northeast; there are also red and white versions.

Identification of the plant

The black currant is a shrub that grows up to five feet tall. It also grows about the same width, making it a rounder shape. The stem is narrow, smooth, and a light brown color. The leaves have three points; the larger leaf in the center is tall and the two on the side slightly jut out. The edges are jagged and the texture is slightly wrinkled. They grow on stems 1-2 inches long in an alternating pattern.

The flower is about a foot long or about the size of a smaller drooping spike. The flowers are sparsely spaced and a pale greenish yellow. The flowers are star shaped, with the petals curling back. The center is prominent and raised like a small cup. The berries are similar to a grape, growing in triangular clusters that droop. The blackcurrent berries tend to be less densely packed together though. The berries are a bit smaller and round, not oval, similar to a black blueberry.

Where to gather it

The plant can be damp areas with fertile soil. It flowers in the spring and fruits in the summer.

How to gather it

Pick the berries off the bush when they are dark purple or black. They stain, so wear protection and old clothes.

How to cook with it

Edible: While the berries can be eaten raw, they are usually processed. They are usually made into some jam like preservatives or wine. See brambles for berry jam recipe.

Author: My grandmother used to treat urinary bladder infections and used black currant, like many do, to treat it. It is high in vitamin C.

DANDELION

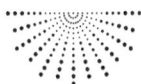

*D*andelions are scientifically known as *Taraxacum officinale*. They are a common weed in lawns, but dandelions are a powerhouse when it comes to foraging.

IDENTIFICATION OF THE PLANT

While most people know what a dandelion looks like, since they are not always in bloom, knowing the whole plant in all seasons is a good idea for foragers. The yellow bloom is iconix, being a half globe with sting like petals that are a dark yellow. This eventually turns into a cottony seed globe that people blow in the wind. The leaves come from a center point at the base of the flower's stem. These leaves are long, up to five inches, and narrow with irregular reverse scalloped pattern, jagged edge. The stem on the flower is circular and hollow. The plant might produce a white sap.

Where to gather it

Dandelions are really common, but they shouldn't be harvested from just anywhere. Because they are considered weeds, it is likely that people are using chemicals to kill them, so make sure that you only harvest from trusted areas. Dandelions like most places but are commonly found in places that are weedy and have been disturbed. You can find them in the summer, but they can be foraged in the spring, fall, and winter.

How to gather it

You can dig up the roots in the spring with the leaves or the fall and winter. This will probably need to be done with a shovel. You can pick the leaves during the spring as it's the best time for a fresher taste. The flower heads can also be harvested at this time.

How to cook with it

Edible: The yellow leaves are edible and can provide vitamin A and K. They are best eaten when they are young because they can quickly turn bitter. They can be used like most greens in salads and even boiled. The flowers are used in salads and for tea. Older plants and roots should be boiled before consumption. Roots can be peeled, boiled, and eaten or roasted and used for a coffee supplement.

Dandelion Flower Cookie

    Time: 17-22 minutes
    Serving Size: 9 servings
    Prep Time: 7 minutes
    Cook Time: 10-15 minutes
    Ingredients:

- 1/2 cup Dandelion flower tops
- 1 cup flower
- 1 cup oatmeal
- ½ cup vegetable oil
- 2 eggs
- 1 tsp vanilla extract

- ½ cup honey

Instructions:

1. Preheat the oven to 375 °F.
2. Clean dandelion heads.
3. In separate bowls, mix the liquid and dry ingredients.
4. Combine the dry ingredients and the dandelion heads into the wet ingredients.
5. Scoop the batter with a spoon onto an oiled tray.
6. Bake for 10-15 minutes.

Author: Dandelions are another flower that people use for its health purposes. They are full of antioxidants, they can help with blood sugar, cholesterol, as well as a diuretic. Because of their health benefits, you should double check to see if it can interfere with any medications you are taking.

My mum used to prepare dandelion salads every day for every meal in spring because, as she liked to say, they detoxed our liver and blood and made us restart fresh into the year's cycle.

DAYLILY

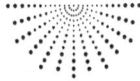

*D*ay lilies are actually not part of the lily family, and the most preferred, when it comes to flavor, is the orange daylily or *Hemerocallis fulva*. While a lily has small leaves that grow horizontally up the stem, a day lily has long leaves that stand the length of the plant. They can also resemble the iris whose leaves are thicker and erect.

IDENTIFICATION OF THE PLANT

There are multiple kinds of daylilies but we will look at the previously mentioned *Hemerocallis fulva*. The plant typically grows up to four feet tall. It has a stem that branches 1-5 flowers. The flowers are about 6 inches wide and 6 inches deep in the shape of a funnel. It branches into 6 petal-looking points. The color is a range of orange shades. A day lily bloom only lasts one day. The leaves, as described by looking at it from above, come from a center point at the ground and are semi-erect. They are four feet and droop around the main flower stem. They are green and narrow, at only about 1-2 inches. The amount can vary, with some looking more or less lush.

Where to gather it

This plant most likely won't be found in the wild, but once planted they can last a long time in gardens and abandoned areas. They like lots of sun. They bloom in the summer but can be harvested in the fall and spring.

How to gather it

Harvest root tubers with a shovel. Harvest only the frim part and replant what you aren't using. You can harvest bubs or blooming flowers by plucking or cutting them off the stem.

How to cook with it

Edible: You can eat the root raw or boiled, similar to potatoes. Cook buds (like green beans) and flowers, either by boiling or frying them, but don't eat it raw. You can also dry flowers and buds and use them as a herb later on. Don't eat the stamen inside.

Daylily Root

Time: 15-30 minutes
Serving Size: 4 servings
Prep Time: 5 minutes
Cook Time: 10-25 minutes
Ingredients:

- 4 cups daylily root
- 2 tbsps butter
- 1 pinch salt and pepper

Instructions:

1. Clean and chop the roots.

2. Bring water and salt to a boil.
3. Add the daylily roots.
4. Boil for 10-25 minutes until tender, and then strain.
5. Add butter, salt, pepper, or dress like your typical potato.

Author: This plant is poisonous to cats and should be used with caution. Some people can be allergic, and it can act like a laxative to some.

EASTERN REDBUD

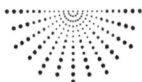

The Eastern Redbud is also known as *Cercis canadensis*.

IDENTIFICATION OF THE PLANT

This is a short tree, usually only up to 30 feet tall. It is ornamental and has a bonsai looking silhouette. With it being flatter on the top, its branches grow wider. Their branches are more organic flowing but not dense. The bark is dark gray and smooth when young, and texturizes as it matures. The leaves are round, heart-shaped, smooth edged, and a medium green color. The leaves are also slightly glossy and alternate patterns. The seed pods look like green beans.

This plant is most identifiable with its show of pink flowers. It flowers mid-spring before the leaves come out. They grow in small clusters out of the branch on 1-2 inch pink stems. There is a round pink bracket at the base of the flower. The petals are pink as well, but more purple toned than the stem. There are 2 petals on one side that sit side by side. On the other side, two petals splay out like wings and, between them, a fifth, slightly darker petal that juts outward like a tongue.

Where to gather it

The plant is often planted for ornamental use but thrives in areas that are moist and sunny like a hill. It blooms mid-spring. Harvest the flower buds before they bloom, and keep an eye on the tree starting early spring.

How to gather it

The tree is short so it is easy to pick the flowers or buds off of the tree when they are in bloom. Since the flower grows in clusters right off the branch, it might be more time consuming than the plants that can be collected off of generalized stems. The seed pods are ready when they are green.

How to cook with it

Edible: The flower is eaten fresh like in a salad but can be fried or made into jams. The buds can be dried and used as a spice, eaten fresh, or cooked. The seeds can be roasted, although it's not very common, or eaten raw like peas.

EASTERN WHITE PINE

The eastern white pine, often just called a white pine, is scientifically known as *Pinus strobus*. If you are a beginning forager, you might not know much about trees and how diverse the category of pine tree is. For example, the red pine has lighter peeling gray to red bark and sparse branches and needles. In general, that makes it not a very appealing tree to look at. For the most part, needled trees are not the first thought for a lot of foragers because they don't produce the same flashy fruit as most deciduous trees. As such, conifer trees can be a blind spot for many foragers. Once they are distinguishable, however, they have the potential to brighten up your kitchen.

IDENTIFICATION OF THE PLANT

The eastern white pine can be a beautiful tree, standing up to 230 feet tall. Its bark is thick and an ashy, medium brown color. It is deeply grooved and textured. While they do typically lose some of the lower branches, they still remain semi-lush. The lower branches can be fairly thick.

The needles on this tree are about three inches long, thin, and dark green. This gives the tree a soft look and feel, they are not prickly like other needles. The new growth tends to be a lighter green and the older needles turn orange and fall off.

The pine cones are long, about half a foot, narrow, and taper slightly to the end. The cone opens, and the fold juts out; they are thick, like certain types of paper, but are also noticeably less dense or woody than other pine cones. The pollen cones stick out like soft white miniature-looking cones that shed an onion-like skin.

Where to gather it

The tree is common and can be found in most wooded areas. They like acidic, well draining soil. Pine nuts are ready late summer. The needles can be harvested any time.

How to gather it

Find a tree that has branches that are low enough to reach. The pine nuts have to be harvested from the pine cone before it opens. You'll know if the pine cone has opened if the flaps are pointed down. If they haven't opened, they will be flimsy.

How to cook with it

Edible: The pine nuts are a common food, often used in pestos. The needles are almost exclusively used to make tea. The bark is used as an emergency food, although it is typically not used anymore. The pine contains chemicals that can be poisonous with excessive use.

EASTERN WHITE PINE CONES

Time: 3-4 weeks
Serving Size: 9 servings
Ingredients:

- 1 bag of pine cones

Instructions:

1. Once you have collected the cones, they need to be dried. They are harvested before they are ripe, so they need to be kept in a bag with airflow like a fabric bag.
2. Place the bag in a sunny area and let them dry for about 3 weeks.
3. Once they are dry, you can hit the bag against a tree a few times to loosen the cones up and separate the pine nut from the cone.
4. Open the bag, checking that each pine cone has been smashed.
5. At the bottom of the bag will be the pine cone flaps mixed in with the pine nuts. Those can be sorted.
6. Most people prefer to toast the nuts. They can be eaten alone or added in recipes for flavor.

Author: Although the pine nuts are edible, the pine tree is usually seen as a very last ditch effort when it comes to foraging for food. This being said, it is a really common tree. For those worried about worst case scenarios, this tree is a comfort to them all. The bark on this tree can be eaten, although it's usually the inner bark that is dried and/or ground into a powder to use as flour.

ENCHANTER'S NIGHTSHADE

*T*he enchanter's nightshade is scientifically known as *Circaea lutetiana.*

IDENTIFICATION OF THE PLANT

The Enchanter's nightshade grows up to about two feet tall. The thin stalks are covered in thin hair, and the bottom of the leaves are slightly toothed. The plant has opposite, simple leaves. The leaves are elongated and oval-to-point in shape. They get smaller the higher they are on the stem.

The flowers are white and bloom in the summer. They have 4 petals in pairs on either side of the center, 3 long stamen, and brackets that are pink in color and also covered in hair that are placed sparsely on the stem. This creates a small bur for a fruit.

Where to gather it

The plant can be found in woodland areas, especially near banks. It blooms in early summer.

How to gather it

Pick berries in mid-summer.

How to cook with it

Edible: No, it is used for medicinal tea.

The tea is made from berries but mostly used as a topical ointment on wounds.

EPAZOTE

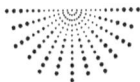

*E*pazote, also known as Mexican tea, is scientifically
known as *Dysphania ambrosioides*. It is poisonous in
large quantities.

IDENTIFICATION OF THE PLANT

This plant grows up to four feet tall. The stem is red. The
leaves are long and narrow, ending in a point, and the edges
are lined with wide, dull points. The flowers are small, green,
and unremarkable. There aren't many irregular branches and
the plant looks rather weedy.

Where to gather it

The plant likes sunny areas with well draining soil.
However, it also likes lots of water. It can be harvested whenever leaves come out in spring to fall.

How to gather it

Harvest the young leaves frequently to encourage more
growth throughout the whole season. If you are not using it
right away, dry it for later use.

How to cook with it

Edible: The leaves and young stems can be used fresh or dried to add flavor. However, it should be added to food at the end of cooking so as not to cook off the flavor. It's wonderful to make tea with it.

Epazote Tea

Time: 2-3 minutes
Serving Size: 1 serving
Prep Time: 1 minute
Cook Time: 1-2 minutes
Ingredients:

- 1 tbsp dried tea leaves
- 1 cup water
- lemon and sugar for taste

Instructions:

1. Bring water to a boil.
2. Add the tea leaves and boil for 1-2 minutes.
3. Remove from heat.
4. Steep 1-5 minutes.
5. Prepare as desired.

Author: One should not ingest too much of this plant because it has a plethora of medicinal qualities, most involving aids of the digestive tract, respiratory, and menstrual issues.

38

EVENING PRIMROSE

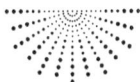

*lso known as *Oenothera biennis*, the flowers of the primrose open at night, hence the name "evening primrose." The evening primrose might be a name that rings a bell for you as it is a common supplement females take to help their premenstrual cycle symptoms. Since this plant is used for medicinal purposes, it is best to not over consume it if possible. There are some primrose that are poisonous and will cause skin irritation if touched, more specifically the german primrose.

IDENTIFICATION OF THE PLANT

The plant can grow up to five feet tall with an erect stem. The leaves angle upward up the stem. They get smaller as they go up, giving it a triangular shape. They are long and thin, reach up to 10 inches, with smooth edges and a red hue on the stem.

It flowers in the second year, at which point leaves will alternate on the stem. The flowers are 1-2 inches wide, a bright yellow, and shaped like a cup. The floret has a small cluster at the top of the flower that has about 3-5 flowers blooming at once.

Where to gather it

The plant likes more open areas, especially places that are weedy. They can be harvested in the spring, fall, and winter.

How to gather it

It is best to harvest this plant's roots in its first year after it has gone dormant in the fall or winter. The leaves can be harvested in the early spring.

How to cook with it

Edible: Yes, roots and leaves.

Peel and boil the roots and use in a recipe. For the leaves, you can eat them raw or boil them to eat like any other greens.

EVENING PRIMROSE ROOT CONDIMENT

Time: 2-3 weeks

Prep Time: 2-3 weeks

Cook Time: 0 minutes

Ingredients:

- 1 cup chopped root, 1 inch in diameter
- garlic cloves, adjust to your taste
- 1 cup white or apple cider vinegar
- 1 tsp rosemary
- optional: 2 tsps sugar

Instructions:

1. Clean and chop the roots.
2. In a jar, combine all of the ingredients and seal with a lid.

3. Let infuse for a few weeks in the fridge.
4. Use roots, garlic, and vinegar as addition to stir fry, sandwiches, salad, and salad dressing.

39

FALSE SOLOMON'S SEAL

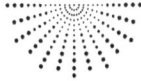

*F*alse Solomon's seal is also called "feathery false lily of the valley" (note: there is another plant called "false lily of the valley," which is also known as the Canadian mayflower). Its scientific name is *Maianthemum racemosum*. The real Solomon's seal is listed in chapter 93. Lily of the valley is poisonous, so it's really important to identify them correctly. The real lily of the valley is much smaller, being only a few inches tall. The 2 or 3 leaves it has grows from the stem and forms around the center stem with flowers. The flowers—small, white, scalloped edged, and bell shaped—latch on a drooping stem. They dangle from small stems towards the ground.

IDENTIFICATION OF THE PLANT

False Solomon's Seal/False lily of the valley is about two feet tall and does not stand erect. The stem is woody. The leaves are long and narrow, at about seven inches long. It has about three parallel main veins running the length of the leaf.

The leaves alternate in two rows on either side of the branch. The edges are smooth and the end comes to a point.

It stands on a branching floret, with many small white flowers in small balls. The fruit are small, white or red berries. They are round and smooth, and the skin is slightly transparent. They can range from dense clusters to only a few.

Where to gather it

The plant can be found in woodland areas, especially near banks with well drained soil. Gather shoots in the spring. Berries are ripe very late in the season, even into the later parts of fall.

How to gather it

Harvest shoots by cutting the young, tender plant before it matures. Pick the berries when they go completely red.

How to cook with it

Edible: Yes, all of it can be eaten.

Young leaves can be eaten, although they aren't preferred by very many foragers. They should only be consumed in moderation as they can have a laxative effect. Young shoots in the spring cook like asparagus. Roots can be boiled and eaten.

See Asparagus for recipe.

Author: Since the berries can aid in digestion, although they might not be the best snack, they can be helpful to have around when in need. This plant should be considered once a forager has a little bit more experience and is confident when it comes to identifying a plant early in the season, especially ones with many look-alikes.

40

FIELD GARLIC

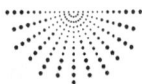

*lso known as wild garlic or *Allium vineale*.

IDENTIFICATION OF THE PLANT

The field garlic grows up to a foot tall. It is centralized and looks like a tall patch of grass. The leaves are round, hollow tubes that end in a point. The leaves might be straight but often have a wave to them. The flower comes up in a separate stem, has a bracket and a cluster of small flowers in a floret called an inflorescence. The flowers are typically light purple, small but erect, and look almost spiked with a round silhouette.

The plant has a strong garlic smell.

Where to gather it

Field garlic is a fairly common plant that can be found in areas that are open from an undense wood to a field. You can find it in the spring through the fall, ready to harvest.

How to gather it

You can harvest the whole plant in the spring or fall,

including the bulb under the ground. You can harvest the green leaves throughout the whole season. Collect flowers when they appear in early summer.

If you are harvesting the bulbs, they can be hung and dried to preserve them.

How to cook with it

Edible: Field garlic can be used as garlic in any dish. Every part of the plant can be used for this flavor. All of it can be thrown on dishes raw or cooked, but the more it is cooked the less strength it has. Since leaves are already more mild, they are typically used raw. The flowers are also typically used raw or as an ornamental look in foods like salads.

Use as a spice.

Author: Adding plants like field garlic, and other strong flavors, can be a great way for new foragers to start building dishes out of foraged material. A lot of other plants can be used as spices too, but having a reliable staple like this can bring so much more potential to substantial, enjoyable, foraged dishes.

41
FRAGRANT SUMAC

*F*ragrant sumac or lemon sumac is scientifically known as *Rhus aromatica*, not meant to be confused with <u>staghorn sumac</u> or poisonous sumac. It can resemble poison ivy; so, when in doubt, don't risk it.

IDENTIFICATION OF THE PLANT

It grows into a spindly bush about five feet tall or stays close to the ground like a vine. The leaves are deeply lobed into three, giving it the look of individual triangle shape as it is narrow at the base and widens at the end. The ends are reversed. They are dark green, thick, and glossy. In the fall, they turn bright red.

The flowers are not very noticeable, growing in clusters at the end of the branches. They are pale yellow, small, and cup shaped. The berries grow in clusters that can be short, slightly pointed, but not as large as the staghorn. They are red, with sparse, fine, white hairs.

Has a lemon scent.

Where to gather it

They are fairly versatile, being able to grow in sun or shade. It is used to rehabilitate areas by giving ground cover. Sumac can be harvested in the summer to early fall.

How to gather it

The berries can be picked off the plant when they are ready in the summer.

How to cook with it

Edible: Yes, the berries are used to make a sour drink that is similar to lemonade.

See <u>Staghorn Sumac</u> for recipe.

42

GARLIC MUSTARD

*G*arlic mustard's scientific name is *Alliaria petiolata*.

IDENTIFICATION OF THE PLANT

The plant grows in its second year up to three feet. Oftentimes because it's an invasive weed, it will be in clusters or patches. The plant is herbaceous, with heart shaped leaves that have large serrations on the edge. It has a small cluster of 5-10 small, white, star-shaped flowers at the top of the stem, and it leaves a garlicky smell.

Where to gather it

Garlic mustard is invasive, but it does best in areas that are slightly shaded. It can be harvested from spring through early summer.

How to gather it

You can harvest the younger plant leaves, flowers, and seeds. In many areas, it's encouraged to pull the whole plant out of the ground because of its invasiveness.

How to cook with it

Edible: Yes, it can be used raw in salads or boiled like spinach. Has a garlic flavor and can be used to add taste to dishes.

FORAGED COOKED MIXED GREENS

Time: 7 minutes
Serving Size: 4 side servings
Prep Time: 2 minutes
Cook Time: 5 minutes
Ingredients:

- 3 cups garlic mustard
- 3 cups chickweed
- 3 cups lambs quarters
- 1 cup dandelion leaves
- field garlic and onion for taste
- 2 tbsps butter
- 2 tbsps water

Instructions:

1. In a wide pan, warm the butter, garlic, and onion.
2. Add a handful of greens one at a time, then add water to the stem.
3. Cook for 5 minutes until tender.

Author: Garlic mustard is really strong when raw but mild when cooked. When the plant is raw, it is better used as a herb. When cooked, it's better used as a leafy green.

43
GINKGO

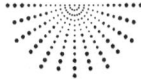

inkgo is a tree, also known as *Ginkgo biloba*. This tree can cause an allergic reaction similar to that of the poison ivy family. Wear gloves, and if you have severe reactions to this family, do not eat it, even if cooked.

IDENTIFICATION OF THE PLANT

Ginkgo trees tend to be fairly large, reaching up to 120 feet. The branches are a little more organic in their shape, sometimes twisting. The bark is thick, light ashy brown, with deep grooves and texture. The flower is not distinct, being drooping, sparse florets, which are indistinguishable flowers. The fruit smells like vomit; they are round, yellow green to light orange, growing in clusters. The fruit is a thin layer of flesh around the nut, which is about an inch long and similar to an almond in appearance.

The easiest way, in my opinion, to identify this tree is the leaves. The leaves are distinct because they have this 1-2 inch green stem and then fan out, like a paper fan at the end. The sides of the fan are flat and the end is rounded out and slightly lobed. The leaves are also thin, watery, and a yellowish green that goes golden yellow in the fall. They grow alternately and sometimes out of clusters.

Where to gather it

The tree likes full sun. They are typically found in parks and other public spaces. They produce fruit in the fall. The leaves can be harvested from spring to fall.

How to gather it

The leaves can be cut or plucked from lower hanging branches. When the fruit is ripe it will fall off the tree. Using gloves you can pick them up. It is recommended you have a very good seal on whatever you are using to transport the fruit due to the smell.

How to cook with it

Edible: Yes, the fruit, the nuts, and the leaves.

Eating too many nuts can be poisonous. As such, try to eat less than five cooked pieces a day. The flavor is sweet to bitter with a hint of cheese. While the fruit can be eaten, most people prefer not to because of the vomit smell.

The leaves are mostly used to make tea.

Prepare Ginkgo Nut

The nut can be eaten alone in small amounts or used in cooking for its taste. **Ingredients:**

- 1 cup Ginkgo fruit

Instructions:

1. Outside, you can remove the thin layer of fruit from the nut, this can be done by hand or by soaking for a few hours in warm water. Be sure to use gloves the whole time.

2. Break the shell with brute force or soften the shell by boiling or roasting them.
3. The shell will crack, similar to a pistachio.
4. You will know the seed is cooked if it turns green or amber, glossy, and chewy.

Author: The leaves of the tree are used medicinally to improve cognitive function of all sorts.

44
GLASSWORT

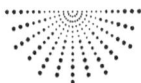

G lasswort is a name for the genus *Salicornia*.

IDENTIFICATION OF THE PLANT

The look of glasworts are fairly unique looking, although fairly area specific. They are small, only reaching a little over a foot in height. They have a shrubby look to them from a distance with its many branches. The branches are stiff and straight. The plant is actually succulent and does not have leaves, just the green branches. The branches have a texture to them that is scale-like, like asparagus. The flowers appear late in the season and are tiny and white out the side of the branches.

Where to gather it

Glasswort grows in salt areas, such as coastlines, salt lakes, and the like. You can harvest from spring to fall.

How to gather it

Pick the most tender tips from the plant's branches.

How to cook with it

Edible: Yes, you can eat the tender tips raw (you'll love how crispy they are). They can also be added to other dishes as a crisp and salty addition. They can also be pickled.

45

GOLDENROD

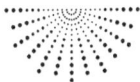

*G*oldenrod is a genus of yellow flowered plants under the scientific name *Solidago*. All varieties of gold-enrod can be edible, but there is a preference for *S. virgaurea*.

IDENTIFICATION OF THE PLANT

On average, the plant gets to about 3-4 feet tall. The plant stands erect and has a hardy stem with slight toothing and a red hue. The leaves tend to be fairly long and narrow in an oval shape with a prominent tip. The leaves are also covered with slight toothing. The edges are uniform and slightly serrated.

The flower spike is on the top of the stem. It is formed by having multiple 1-2 inch stems that protrude from the main stem, creating a triangle shaped spike. Each stem is lined with multiple, very small, yellow flowers.

Where to gather it

Goldenrod blooms in the late summer and into the fall. Goldenrod is a common weed and can be found in typical weedy places such as ditches and other open places, especially ones that have been disturbed but aren't excessively damp.

How to gather it

There is fine toothing on the plant so be cautious. Harvest flowers when in season. Harvest the leaves right before the plant flowers. If you are planning on harvesting the flowers to dry, make sure the plant is still young enough that it won't go to seed.

How to cook with it

Edible: The flowers can be eaten raw or dried for teas. The leaves can be eaten raw and dried for teas as well, but it also functions as a leafy green and can be treated like spinach.

Author: Goldenrod has a bad reputation for causing the pollen for seasonal allergies, but it's actually ragweed that is responsible for hayfever. This being said, people who have seasonal allergies should be careful consuming anything that has pollen.

GOUTWEED

G outweed is known by many names such as Ground elder, bishop's weed, and more, but it's all under the scientific name *Aegopodium podagraria*.

IDENTIFICATION OF THE PLANT

The plant is not very tall. It grows typically in small clusters from 1-3 feet tall. The leaves give the plant a fair amount of foliage. The stems stand straight and are hollow with a texture of grooves. The leaves are elongated oval shaped with jagged edges packed in groups of 5.

Goutweed looks very similar to wild carrot because its flower blooms in the spring and summer. The flower is a cluttered floret umbel that fans out flat in a roundish shape that's 4-6 inches wide. It consists of small white flowers.

Where to gather it

This plant is considered an invasive plant, meaning that it can go pretty much anywhere. Like most weeds though, they like places that have been disturbed or abandoned. It grows from the spring to the fall.

How to gather it

Gather only the young parts of the plant and cut back the rest. This plant is hearty and when it's cut back, it will grow more young leaves for further foraging.

How to cook with it

Edible: The young leaves can be used in the same way as spinach: in salad or cooked. It has a peppery taste.

Nettle and Goutweet Soup

Time: 15-20 minutes

Serving Size: 4 servings

Prep Time: 5 minutes

Cook Time: 10-15 minutes

Ingredients:

- ½ cup Goutweed
- ½ cup Nettle
- ¼ cup dandelion
- 1 liter soup broth of choice
- ½ cup cream
- 1 onion
- 1 clove garlic
- 3 tbsps butter or other oil
- 1 tbsp flour
- for taste: parley, salt, and pepper

Instructions:

1. Clean all of your produce.
2. Chop everything finely.
3. In a pot, add the oil over medium heat to cook the garlic and onion.
4. Add in flour and stir.

5. Add broth, dandelion, nettle, and goutweed.
6. Bring to a boil, then remove from the heat.
7. In a blender, pour the soup and puree into a smooth texture.
8. Add in cream and spices and puree for a few seconds.

Author: The flower is used as a laxative, so be careful to only use the young parts of the plant. Similar plants are a young elder tree (hence the name) and Dog mercy, which has hair on the stem and leaves.

GRAPE

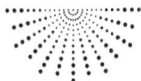

The grape is a fruit that grows from a grape vine. The genus of the grape is called *Vitis* and there are many species apart of this genus, but the ones that are popular to this area are *V. labrusca* (fox grape), *V. aestivalis* (summer grape), *V. vulpina* (frost grape), and *V. vinifera* (common grape).

IDENTIFICATION OF THE PLANT

The grape vine is firstly identified by the form of its vine. Like most vines, this can mean that if it has somewhere to climb, it can climb up trees or across fences, but it can also crawl across the forest floor if it needs to. The older stem of the vine becomes woody, and, depending on the variety, it can become a few inches thick and create a thin, peeling bark. The younger parts of the plant are red or green, and while still hardy, it's got a smooth texture.

The leaves of the grape vine alternate, and while they vary, they do have similar looks, each being round with three or five main lobes with a main vein that branches to each and larger jagged edges.

The fruit of the grape can vary between the species. Just looking at the differences from the four species listed above, the fox grape is typically red and dusty, larger, and round. The summer grape is smaller and solid green to a solid purple color. The frost grape can range in size but is more of a bluish tone, and the common grape tends to have a more dense, larger cluster of berries.

Where to gather it

Grape vines like places with water and places to climb. This typically means that you can find them near wooded lakes and the like. The leaves are harvested in the spring and the fruit is harvested throughout the late summer and into the fall. The frost grape specifically should be harvested after the frost, to which it gets its name, for the sweet flavor.

How to gather it

The leaves should be young but fully grown, and they can be cut or plucked from the plant. The fruit can be cut off in its triangular clusters when ready to eat. You can taste test to see if the grape is desirable. If you are looking for grapes that are sweet enough to eat on their own, this is especially recommended since many wild grapes are tart.

How to cook with it

Edible: The fruits can be eaten and used like any other store bought grape. Typically since it's not as desirable in flavor, they are typically used in everything from deserts to preserves or raisins. The leaves can be boiled and eaten like other leafy greens.

WILD GRAPE JUICE

This recipe makes 3 cups of concentrate for 1 gallon of water or 3 tbsps for 1 cup.

Time: 30 minutes
Serving Size: 4 servings
Prep Time: 5 minutes
Cook Time: 25 minutes
Ingredients:

- 1-2 galleons grapes
- 3 cups sugar
- 2 cups water

Instructions:

1. Clean and remove grapes from stems.
2. Start pressing the grapes, squeezing the liquid from the flesh.
3. Once this is done, strain with cloth, squeezing as much juice out as possible.
4. Let sit in refrigeration overnight to let material settle.
5. The next day, gently pour off clean juice, throwing out the bottom impurities. Yield of juice 3-5 cups.
6. In a large pot, mix water and sugar over medium to high heat to create a syrup.
7. Add in grape juice.
8. Taste test and add in sugar if needed.
9. Bring to almost a boil, remove from heat.
10. Pour into sterilized jars, seal, and label OR Freeze (not in glass).

Author: Grapes, like other wild fruit, should be picked with caution. There are a lot of wild berries you can eat but also many you cannot. The grape's most similar look alike is the moonseed (*Menispermum canadense*).

48

GREENBRIER

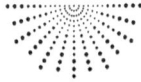

*G*reenbrier usually refers to a genus, but in this chapter we are going to focus on common greenbrier. This species' scientific name is *Smilax rotundifolia*.

IDENTIFICATION OF THE PLANT

This plant is a vine, meaning that you might find it climbing something taller or you might find it creeping across the ground as it extends about 20 feet. It has thorns and creeps with tendrils. The older growth can become woody, so its stems can range from a herbaceous green to hard and brown. The leaves are a dark glossy color and the bottom of them is a paler green.

The flower usually branches out in clusters off of a stem. They start as a small green oval and bloom into small, star-shaped, white flowers. The berries are very similar to a blueberry but form a densely packed, round cluster.

Where to gather it

The plant can be harvested in the summer but it's typically sought out in the spring. It can be found in places with openings like ditches, open woods, banks, and the like.

How to gather it

The plant is easy to harvest since it is herbaceous and within arms reach, meaning you will only need some scissors. The trick is to harvest the part of the plant that is still young, or at least fresh. Include the tendrils, leaves, and new growth. Be careful because of the thorns.

How to cook with it

Edible: Yes, the younger parts of veins and leaves can be eaten raw, by itself, or part of a salad. Its shoots can also be cooked like asparagus, while the rest can be treated like other leafy greens. The roots have a gel that can be used for thickening.

See Asparagus for recipe.

Author: There is not much information on the edibility of the fruit; some sources say they are edible through the winter, but not substantial.

49
GROUND CHERRY

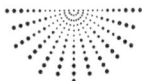

*T*he ground cherry is known scientifically as the *Physalis* genus. This plant is mistaken for gooseberries and even called the cape gooseberry on occasion. Gooseberries do not have a lantern and look more grape-like than tomato-like.

IDENTIFICATION OF THE PLANT

The plant is native to the Americas, and therefore it's important to recognize the variety of the genus. The species can range from a few inches tall to a few feet tall. It is typically erect if the plant is of the taller variety.

The stem and branches can have a purple tone to it. The plant greatly resembles the tomato plant in its silhouette and can almost appear like its vining. The vining look is especially noticeable in the shorter species that grow out like ground cover instead of up. The leaves are also similar, being alternate, simple, and oval to a point, with some varieties having lobed or scalloped edges. The branches, stem, and leaves are covered in a soft fuzz. The flower is quite different from tomatoes, although it's a shallow trumpet shape with a dark brownish on the lower or inside portion of the petals and a very pale yellow outer edge. The center of the flower is a solid point, no prominent stigma.

Unlike many fruit bearing plants, you can't identify a ground cherry from its fruit because of its lantern-like husk. The fruit on the inside is very similar to a cherry tomato, and typically a yellow to orange color. The husk on the outside is thin like a membrane and green to light brown. It is round in shape around the base and comes to a point at the bottom of the fruit. It has seams from the top to the bottom point of the lantern, giving it panels.

Where to gather it

They like full sun and are not fond of the cold. They like places like a less dense wood because of the ability for it to be open and also slightly sheltered. The fruit is ready around late summer and should be harvested before the cold weather hits.

How to gather it

The fruit of the groundcherry can be a little more difficult for some foragers because it requires some strategy. The fruit of the plant typically falls before it is ready to be eaten, so it needs to be harvested beforehand. Once harvested, the fruit needs to be stored to allow it to ripen for up to a few weeks to get the sweet flavor. Unripe fruit can be poisonous. Store in a dry place.

How to cook with it

Edible: Yes, this plant can be eaten raw, used in deserts, or preserved. The fruit must be ripe and removed from its papery husk before eating or cooking. A good indicator of when the plant has ripened is when the husk is looking dried out instead of green. The flavor of ground cherries can be sweet and used like dessert, or more like its family member the tomato.

. . .

GROUND CHERRY SALSA

This recipe makes 1 large jar of salsa. This is not a preserved recipe.

Time: 24-48 hours
Prep Time: 24-48 hours
Cook Time: 0 minutes
Ingredients:

- 1 cup ground cherries
- ½ cup onion
- 6 tbsps chopped tomatoes
- 2 tbsps lemon or lime juice
- 1 clove garlic
- 1 tbsp Olive oil
- optional: ¼ cup jalapenos for heat
- optional: cilantro and other spices for your taste

Instructions:

1. Clean your produce.
2. Remove ground cherries from husks.
3. Chop all of the ingredients into fine chunks. If you prefer a chunkier salsa, chop the pieces slightly bigger.
4. In a bowel, add all of the ingredients together.
5. At this point you can let it rest for a day to absorb the flavors.
6. OR If you like very smooth salsa, you can add to a blender with a few tbsps of vinegar. This will keep the salsa preserved for longer in the fridge.

Author: Ground cherries are a part of the nightshade

family. Nightshade is typically referred to when speaking of the poisonous varieties of the genus, but the genus *Solanum* is the umbrella for tomatoes as well. Plants like 'bittersweet', or *Solanum dulcamara*, grow in the Northeast and are part of the same family. This should not be an issue as long as the forager identifies the plant before foraging.

50

HAWTHORN

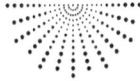

*H*awthorn is the common name for a genus of plants that can take the shape of a small tree or shrub. The scientific name for this genus is *Crataegus*. There is another plant that goes by the common name hawthorn or Indian hawthorn with the scientific name of *rhaphiolepis*. This genus is common in Asia and is used for ornamental use, even being used for bonsai and sometimes bear fruit. The fruit is blue and the leaves are smooth edged, so there shouldn't be confusion beyond the name. The species *C. monogyna* is the most used for food, but all hawthorn berries are edible.

A common look-alike to the hawthorn is the blackthorn (*Prunus spinosa*), but once again the fruit is a dark blue, resembling a blueberry, and blooms later than the hawthorn.

PHYSICAL IDENTIFICATION OF THE PLANT

Typically a small tree, growing up to 40 feet for some species but can be as small as five feet for a fully grown shrub. The leaves alternate and are simple with variations of different points and usually with some level of jaggedness on the edge. They are deciduous, but some variations might be more hardy and therefore slightly evergreen, but not completely. This means that some leaves are more watery and yellow-green and some are darker and glossy.

The bark is gray and smooth when young. As it ages, it gets more texture for most species. Many of the variations have very stiff, straight branches.

The lucky thing about plants with foragable fruit, the fruit tends to be the easiest thing to identify, unlike plants that need to be harvested before their identifiable attributes are on display. Similarly, the fruit of most hawthorn varieties tend to be very similar. The fruit is similar to a cherry in size, but it is typically slightly more elongated like an oval than completely round and it also has a pit in the center. The color is a very bright red, and it hangs from a 1-2 inch stem. The bottom of the fruit has a small crown similar to a blue-berry. In some cases, the fruit on some varieties can range from black to light yellow, but this is species specific.

Where to gather it

The fruit of the hawthorn genus is ready to pick in the fall, but the wide variety of hawthorns can be found in different environments. The fruit is not quick to rot though and may be reliable for weeks to even months. You might find hawthorn varieties near bodies of water, some small banks, some prefer the coast, or even on slopes like hills and mountains. Since they are all edible, finding the hawthorn that grows best in your area is the best bet.

How to gather it

The plant only grows into a small tree. So that when you find a hawthorn tree or shrub, you should be able to reach the fruit without a problem. You should be able to pick the fruit easily, but it may be in your best interest to taste test the fruit first to make sure it is to your liking and at the best ripeness. Since the fruit can stay on the plant for a while there is no rush. Be careful of thorns.

How to cook with it

Edible: This plant can be eaten as a fresh fruit, put in pies, or preserves. The fruit can also be used to make tea. It is likened to rosehips and can range from sweeter to very tart.

HAWTHORN KETCHUP

The ketchup is good for 1 year.

Time: 45-50 minutes

Prep Time: 35 minutes

Cook Time: 10-15 minutes

Ingredients:

- 2 cups berries
- 1 ¼ cup apple cider vinegar
- 1 ¼ cup water
- ½ sugar
- pinch of salt and pepper

Instructions:

1. Clean the berries and remove any stems.
2. Add to a pot with water and vinegar for an half an hour until skins come off.
3. Use a strainer to remove pits and other impurities.
4. In a clean pan, add sugar.
5. Put on low heat, wait until sugar is dissolved.
6. Bring to a boil, simmer until it starts to thicken (up to 10 minutes).
7. Add your spices
8. Put in a sterilized jar(s), seal lid, and label.

Note: Once the jars are sealed, it's recommended to take the ring off of the jar.

Author: Besides being a genus that is conveniently all edible, the hawthorn also has a long history in myth and lore. While this changes from culture to culture, it's comforting to know that many of our ancestors had a relationship with this plant. It has also been used as an herbal aid for heart health and digestion. Because of these effects, it's recommended to not overindulge.

51
HENBIT

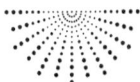

*H*enbit is a deadnettle also known as *Lamium amplexicaule*, also very similar to *L. purpureum*. Nettles are a large group of flowers, some more or less friendly. See chapter 73 for more on nettles. The term deadnettle can sound ominous, but it actually is in reference to nettles that sting and create a rash or irritation. Dead nettle refers to the fact that the nettles are dead, meaning that they don't sting. They can look similar so be careful.

IDENTIFICATION OF THE PLANT

Henbit is a small herbaceous plant that reaches a little over a foot tall and is erect and square. The stem has fine hair and the leaves are heart shaped, at about 1.5 inches long. They come off the stem in an uniform opposite pattern, but in pairs on either side like a ladder. The leaves have a wrinkle-like texture. The edges are loved and have large serrations like points.

The flowers are a magenta pink. They grow right out of the stem. They are long and tubular and open at the end to a non-symmetrical formation. Two petals open out, away from the plant, and one petal on the other side is pointed up and over like a hood.

Where to gather it

Can be found in common weedy areas. It likes more open areas like fields. Young shoots are harvested in the spring. Flowers very early in the spring.

How to gather it

Harvest shoots by cutting the young, tender plant before it matures. Be aware of how early the plant flowers need to be harvested very early in the year.

How to cook with it

Edible: Yes, young leaves, shoots, stems, and flowers.

The edible parts can be eaten raw or cooked, used in the same way as other leafy greens like spinach. The plant is like a celery flavor with a slightly peppery taste.

Author: Deadnettles are also used for medicinal purposes, particularly aiding in digestion as a laxative.

HICKORY

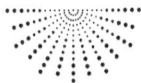

ickory refers to a group of trees under the genus *Carya*. Not all hickory nuts are edible. Mockernut hickory is one of the edible types that is native to the Northeast. It is scientifically named *C. tomentosa*.

IDENTIFICATION OF THE PLANT

The tree grows to about 60 feet tall. The bark is dark gray with groves that give narrow overlapping stripes. Leaves are compound opposite. It's leaflets have a pointed end that is oval in shape and narrows towards the base with parallel veining.

Flowers hang in three inch narrow catkin clusters. The fruit is a hard nut the size of a gold ball with a green outer layer. The nut inside is a little larger than an almond and oval.

Where to gather it

Likes areas that have sloped ground like hills and the like. Nuts are ready to pick in the fall.

How to gather it

Harvest the nuts by letting them fall. You will know they are ready and falling because the leaves will start to fall at the same time. Pick the ones off the ground or, if possible, place a tarp to catch them.

How to cook with it

Edible: Yes, the nut.

Choose nuts that are intact with no bug holes. The shells are thick, so you'll need to use a hammer to crack them open. Be sure to pick a location that can handle the process. Once you've cracked the nuts and loosened the flesh, add them to a pot of boiling water. The flesh will float to the top.

5 3

HIGHBUSH CRANBERRY

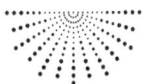

*he highbush cranberry is actually not a cranberry, it is just the common name for *Viburnum trilobum*. It is more closely related to the guelder-rose. This being said, the berry that comes with the plant looks and tastes very similar and is even ready at the similar time of year. This is one of those similar plants that are genuinely similar; however, this is often not the case. It is still important to accurately identify before you forage and eat. There are plenty of other berries that look similar and are not edible. You should never get too comfortable eating strange berries.

IDENTIFICATION OF THE PLANT

The shrub gets up to 20 feet tall in some places. It can grow into a more sparse looking, strangely bush, or remain dense and round looking. The branches have a reddish-yellow color when young and end up gray, but still smooth, as it ages. The leaves are green and have three main points, with a main margin for each and some wider serration on the edges as well. It also has slight hair underneath.

The flower is a round, flat top floret with many white flowers. The flowers have five petals and are in a star shape, but unlike other flowers that share similarities, they are slightly bigger and the petals are rounder and wider. The floret doesn't typically bloom all at the same time, having a few flowers in each cluster at once. The florets are only about a few inches wide, but the plant usually produces many of them.

The berries droop in flat, wide clusters, ranging from a few to many, at about half an inch long. They are bright red and oval. Unlike the cranberry, they are more translucent instead of a solid color. They have a flat pit inside.

Where to gather it

The plant likes moist areas, lowlands, banks, and the like. The fruit is on the plant in the summer but is ready to harvest in the autumn or throughout the winter.

How to gather it

The berries can be picked off of the bush. Unlike the cranberry, the bush is tall and you won't be kneeling to collect. Since they like wet areas, make sure you wear water-proof boots and are safe near water. Taste test the berry; if it is not sweet in fall and winter, it might be the guelder-rose, not the highbush berry.

How to cook with it

Edible: The ripe berry tastes similar to the cranberry. It can be eaten raw, but it can be very sour, so it is typically made into jam. See Cranberry for recipe.

54

HONEWORT

*H*onewort, scientifically known as *Cryptotaenia canadensis*, has some very close look-alikes, many of which are not edible, since it is part of the carrot family. This plant, like the milkweed, is home to a butterfly, the black Swallowtail.

IDENTIFICATION OF THE PLANT

The plant stands up to three feet but is usually shorter. The stem is thin, green, and smooth. The leaves are compound in groups of three and each individual leaf has three lobes with edges that are slightly jagged. The texture is slightly wrinkled and four inches long, getting smaller up the stem.

The flower is a sparse looking floret with very small flowers off of thin branching stems. Individual flowers are not much bigger than the stem itself. They are white with five curling petals.

Where to gather it

The honewort likes moist conditions, as such it is typically found in woodlands. Blooms early summer.

How to gather it

Harvest shoots by cutting the young, tender plant before it matures.

How to cook with it

Edible: Yes, young plant and root.

The stems and leaves can be boiled like spinach or raw for a desired flavor that is mild, bitter, and earthy. Roots can be cooked like any starchy root.

55
HOPNISS, GROUNDNUT

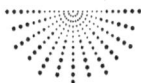

The Groundnut goes by many names—like the potato bean, hopniss, etc.—but is scientifically known as *Apios americana*. There is another plant that goes by the name groundnut. While it's a family of flowering plants, the hopniss plant is a vine. The plant hosts the larva of a butterfly called the silver skipper.

IDENTIFICATION OF THE PLANT

The groundnut is a crawling vine that can grow about 20 feet long. It has a thin green stem and compound leaves. The six leaves are leaflets in opposite pairs with the seventh leaf at the point. They are typically longer and thin but can be more round like a basic leaf. The point is prominent.

The flowers are clusters in a short, dense, almost round, triangular shape about five inches long. The dark pink flowers themselves look sort of unusual with two round petals that open like cupped hands. From the side they look like oysters. The center is like a sideways disk that sticks out like a tongue between them. The root is round like a potato, although much smaller. The fruit is a pod about five inches long that looks like dark brown/green beans.

Where to gather it

The plant is a vine, so you might notice it crawling on the ground or climbing up other plants and such. They like damp ground and are found near the banks and edges of water. Recommended harvest is late fall or early spring, but they can be picked all year.

How to gather it

Since the plant is around banks and such, the ground might be easier to work with; however, still take a shovel, or better yet a garden fork, to dig and sift for the roots.

How to cook with it

Edible: Yes, the seeds and roots are eaten.

While it isn't actually a root, for all intents and purposes as it's a tuber underground, its root is not only edible but delicious. They are cooked like potatoes and eaten as starch. The fruit (bean pods) are also cooked and eaten. The seeds, once removed from the inside of the pod, can be eaten as a legume or bean. The flowers and stems are also edible but not a common snack.

CARAMELIZED GROUNDNUT

Time: 30 minutes

Serving Size: 4 servings

Prep Time: 5 minutes

Cook Time: 25 minutes

Ingredients:

- 2 cups groundnut root
- 2 cups broth
- pinch of salt
- 1 tbsp butter

- 1 tbsp honey

Instructions:

1. Clean, peel, and chop the root.
2. In a pan, add the root, broth, and salt, and bring to a boil.
3. Turn down to a simmer and cook until the root is tender as you would a potato. This should take about 10-15 minutes. The broth should greatly reduce.
4. Turn heat back up, add butter and honey until a sauce is formed.
5. Add seasonings of choice.

Author: A great thing to note about the root is that they are able to store like potatoes. If they are kept in a cool, humid place like a cellar, they can last for quite a while. This is great for winter preparations. While many jams and such can be used for the winter, having a starch like this can be a game changer for those who choose or need to live off of the land.

A really good friend of mine always invites me to eat hopniss. I feel so wild. I think of the Dakota tribe eating them like potatoes. Our meal is like a ritual—we eat on the ground, with our hands, and somehow we always end up with spiritual talks. Such a great time!

JAPANESE KNOTWEED

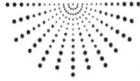

*apanese knotweed is scientifically known as *Reynoutria japonica* or *Polygonum cuspidatum*. It is not recommended to have this plant near buildings because it will grow through them and cause damage.

IDENTIFICATION OF THE PLANT

The plant can get fairly tall, up to 10 feet and erect. The stem is hollow, smooth, branching, and sometimes has a red hue. The leaves are alternating and heart shaped to basic oval-pointed. They are up to five inches long. The flowers are on about five inch long stems that create an erect spike. The veining tapers the leaf slightly. The spike is narrow and the individual flowers are very small and sparse. There are many spikes that line the branches and are thick in texture. Fruit is a small, papery disk with a seed inside.

Shoots look like asparagus with leaves, not scales, on top and bamboo-like seams.

Where to gather it

The plant can be found in weedy areas, like places that

have been abandoned. It is considered a weed. Harvest young shoots in mid-spring.

How to gather it

Harvest shoots by cutting the young, tender plant before it matures. It can help to find the dead plant from the year previous to both spot and identify the plant. It's best eaten if the shoots are under a foot tall.

How to cook with it

Edible: Yes, the young shoots can be eaten raw or cooked. It tastes sour or tart. Can be used in sweet dishes like rhubarb or in savory dishes like asparagus.

Japanese Knotweed Relish

Time: 45 minutes
Prep Time: 5 minutes
Cook Time: 40 minutes
Ingredients:

- 5 cups Knotweed shoots
- 5 cups onion
- 3 cups cider or white vinegar
- 3 cups brown sugar
- 1 tsp each of your choice: cinnamon, black pepper, cloves, ginger, and garlic

Instructions:

1. Clean and chop knotweed shoots.
2. In a pot on medium heat, add the onions (garlic and ginger if using) and soften a bit.
3. Add sugar and vinegar.
4. Turn up heat.
5. Add ¾ knotweed and other spices.

6. Bring to a simmer, lower heat, and cook uncovered for 30 minutes.
7. When thick, add the rest of the knotweed, cook for 10 minutes, then remove from heat.
8. Seal in sterilized jars and label.

Author: Japanese knotweed is used in herbal medicine for many treatments. The whole plant is used to treat respiratory, skin, and oral issues.

JERUSALEM ARTICHOKE

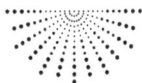

*J*erusalem artichoke has many names, you might know it as sunchoke or wild sunflower among others, but it is scientifically known as *Helianthus tuberosus*. While this plant is useful and has a lovely flower, it can become invasive. If you are considering planting it, keep an eye out to ensure that it does not over-take your garden. It is a part of the sunflower family but does not produce the seeds like many of them do.

IDENTIFICATION OF THE PLANT

The plant can get really tall, like many of the sunflower family, at up to about 10 feet. The leaves are opposite at the bottom and alternate up the stem. The leaves also get smaller up the plant, they are up to 12 inches long, with a rough hair. They are long and narrow in shape to a point with smooth edges.

The flower is shaped like a daisy but has a smaller center and longer, slightly wide petals. The flowers can grow up to eight inches wide. They are a bright golden yellow. The flowers can seem like they are solitary because of their shape and size, but technically they are part of a floret that branches up to about ten flowers at the end of the stem. The rubbers are tear-drop shaped and pointed down.

Where to gather it

Likes places with high moisture in the ground. Found in weedy areas like abandoned places or thickets. Harvest in the fall.

How to gather it

In the fall after the first frost, they can be dug up throughout the winter as long as the ground is not frozen. However, locating them might be hard without identifying tall stems or flowers.

How to cook with it

Edible: Yes, the root.

The tubers can be eaten raw or cooked like other root vegetables. They are sometimes pickled.

Author: I go crazy for it! It tastes so sweet and it has the texture of a potato but the taste of an artichoke. When topped with fresh parsley, it is honestly a journey for your taste buds!

JEWELWEED

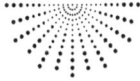

*J*ewelweed is also known as spotted touch-me-not, orange jewelweed, and more, but is scientifically known as *Impatiens capensis*. The stems and leaves can be boiled to help relieve poison ivy rash.

This plant is called a touch-me-not because the seed pods can burst, shooting out seeds when touched.

IDENTIFICATION OF THE PLANT

The jewelweed grows up to five feet tall and is herbaceous. The stems are branched and smooth with high water content. The leaves alternate and grow up to five inches long. The leaves are narrow at the base and widen at the top, while the end is a rounded point. They are smooth and have a long stem. The plant is not overly dense or sparse.

It can bloom from spring to fall. The flowers are orange, sometimes with spots of darker orange. They dangle from a stem but open horizontally. The flowers are unique in shape; they are trumpet shaped but with a long narrow tube. The flower is not symmetrical, having one or two large darker petals on the bottom and a smaller, lighter petal hooding the top.

Where to gather it

The plant can be found in wet solids near bodies of waters like ponds or lakes.

How to gather it

Harvest shoots by cutting the young, tender plant before it matures.

How to cook with it

Edible: Yes, although the young shoots, flowers, and seeds should be eaten in moderation. Boil the leaves twice in new water before eating.

The plant is more commonly used as a medicinal herb to be used on the skin raw or made into other remedies.

JUNEBERRY

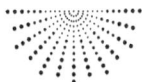

*J*uneberry refers to a genus of plant that can be trees or shrubs, scientifically called *Amelanchier*. They might also be called shadbush or wood, sugar plums, serviceberry, and more. Many, if not all, of the genus is edible, but the downy or saskatoon berry are some of the most common. They go by *A. arborea* and *A. alnifolia* respectively. Downy berries are more common in the East.

IDENTIFICATION OF THE PLANT

The plant can be a shrub or small tree, meaning it can be up to 40 feet tall. The bark is smooth, but the trunk can be twisted or furrowed, although it is typically more narrow. It is a gray color. The leaves are a basic pointed oval shape and a yellow green, growing up to four inches long. A muted red in the fall. They are in an alternate pattern.

The white flowers grow in clusters that cover the tree in the spring. The flowers are star shaped but the petals are long and narrow to prevent it from touching the hole at the center. Prominent stamen comes out of it. The berries are a mix of bright to dark red. They are almost identical to a blueberry, but the crown is red as well. They cluster off 1-2 inch stems in a spaced out group of 10 or so at the bottom few inches of the branches.

Where to gather it

Because of the variety in the genus, there is a variation on where they grow, although they typically grow in areas that have high moisture and a sloping ground. This can range from mountain-like areas to swamps. Berries are ripe in midsummer.

How to gather it

Harvest the berries by picking them off of the bush when they are ripe. You may need a step ladder.

How to cook with it

Edible: The berries can be eaten raw or cooked. They can also be made into jams, used similarly to bramble fruit.

JUNIPER

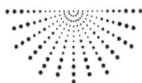

*J*uniper is a genus scientifically named *Juniperus*. They can take the shape of a shrub or tree. Only a few of the junipers are considered edible. They are not native to the Northeast. One of the most popular junipers is the california juniper and it, like the others, typically grows in the west. This being said, there are junipers in the Northeast, but they are not as favorable.

J. communis, or the common juniper, is used in cooking, but usually more for flavor. This will be the juniper talked about in this chapter. Be cautious eating other species unless you have identified it as an edible species.

Juniper berries are often confused for capers. Capers are similar looking to juniper berries but come off of the caper bush. Juniper berries can be pickled in a similar manner though.

Pregnant women should not eat this plant.

IDENTIFICATION OF THE PLANT

The common juniper is typically a low shrub. Junipers

can be dense and used for hedges, but when natural the leaves and branches can spread out into clumps. Without maintenance they can become less appealing because of the rather bare inside branches. They can spread outwards instead of up, creating a flat, disk-like silhouette. This is a coniferous plant, meaning it has needles. The needles are short and prickly. The younger needles might be more green, but the older needles often have a bluish look to them, but not quite as dusty blue as the blue spruce. Some juniper needles appeared to be scaled. The bark is a reddish brown and slightly grooved.

The berries look like blueberries but smaller. They grow in small clusters off of the branches. The cones are very similar looking but not edible. They are more densely packed but slightly scaled and a dusty pale green. Before they mature, the cones are small and pale yellow.

Junipers are sometimes avoided in planting because of their smell, which smells distinctly like cat urine.

Where to gather it

The plant can be found in areas with well draining soil, preferably sandy. Berries are ready in the fall. It doesn't fruit the first year.

How to gather it

Harvest berries by picking them off the shrub when they are blue.

How to cook with it

Edible: The berries are not usually eaten raw because they are bitter. Instead, they are dried and used to season food, mostly meat and alcohol like beer and gin. **Do not eat the cones or leaves.**

PICKLED JUNIPER BERRIES

This recipe is focused on pickled juniper berries that can

be added to other food as a seasoning. If you are creating a pickling mix, do not use more than a tablespoon of the berries.

Time: 25 minutes
Serving Size: 4 servings
Prep Time: 5 minutes
Cook Time: 20 minutes
Ingredients:

- 1/4 cup juniper berries
- 1 cup pickling vinegar
- 1 tbsp salt
- 1 cup water
- 1 tbsp sugar
- spices of choice: garlic, mustard, onion, or pepper
- (if creating picked mix instead, beets and onions are recommended)

Instructions:

1. Wash juniper berries and other produce.
2. In a pot, bring water, vinegar, salt and sugar to a boil until the sugar is dissolved.
3. While the brine is heating, in sterilized jars (heat proof) add a mixture of spices and the berries. Leave room at top, and the berries should be fully submerged.
4. Pour hot brine over berries in the jars.
5. With heat protecting gloves on, tap or shift jars to make sure the brine reaches all the berries and there are no air bubbles.
6. Put on the lid and add to the fridge.

Note: these are not sealed. You can seal them if you want

to for longer preservation. If not, they'll last at least three months in the fridge.

Author: The juniper is used in medicinal medicine to treat arthritis, diabetes, and infection. Juniper alcohol is sometimes used, but theriit bark is also used.

Junipers are sometimes the evil look-alike. Cedar trees are often more desirable. The false cedar is actually a juniper, which makes it all the more confusing. The Red Eastern Cedar leaves, berries, and wood are used in herbal medicine.

LADY'S THUMB

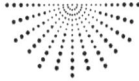

*L*ady's thumb is also called redshank or *Persicaria maculosa*.

IDENTIFICATION OF THE PLANT

Lady's thumb is a common weed, it even looks weedy. It grows up to three feet tall, standing erect. Prominent joints at the branch hinge, stem is green, and hued with red. Leaves are alternate. They are best identified by the dark marks on the leaves. They are patches that look like the plant has been damaged. It gets its name from this, claiming that it looks like a lady's fingerprint. The leaves are simple and elongated with a prominent point. They are light green with a main margin and parallel veining.

The flower is a narrow inflorescence on a stem. The floret is a straight spike like the shape of a hotdog. The individual flowers appear to be closed in an oval shape.

Where to gather it

Flowers in summer. Since it is a weed, it adapts to wet

and dry areas. It is best found in disturbed, overgrown areas. Blooms through the summer.

How to gather it

The young leaves and shoots can be harvested in spring. The flowers can be picked when in bloom.

How to cook with it

Edible: The young greens can be eaten raw or cooked. The greens are similar to lettuce or spinach. The leaves and flowers are dried for tea. The flower can be eaten fresh in salads or as garnish.

Author: Lady's thumb is also used as a medicinal treatment against the poison ivy rash and bug stings.

62

LAMBS QUARTERS

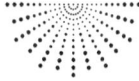

ambs quarters is also known as white goosefoot and scientifically known as *Chenopodium album*. If you have ever weeded a garden, you probably know lambs quarters, if not by name then by annoyance. Lambs quarters is a very common weed, but it has started to revive its reputation due to its nutritious value of fiber, protein, and vitamins.

IDENTIFICATION OF THE PLANT

Lambs quarters can grow up to three feet. When it's younger it is more erect, but as it gets taller, it can droop slightly. The leaves are watery and green with a dusty white cast, much paler on the underside. The shape of them on the younger plant is an equal sided, rounded triangle with very scalloped edges. Newer, smaller leaves growing from the top, full sized, can be up to four inches long. As the plant gets taller, the leaves become longer and more narrow. They are toothed along the margin. As the plant gets older and branches, it looks more weedy.

The top part of the plant branches off in a few about three inch branches to create a narrow, sparse spike. The little branches can start up to over a foot from the top of the plant. The flowers are very small, round, green, and cluster in small, round florets/inflorescence. There clusters of flowers line the three inch branches. It does not look like a flower, it looks very weedy.

Where to gather it

This plant is a weed, so it's very common in most places, especially over ground areas. It is best to avoid plants in areas that can be contaminated. Best to harvest the greens in the spring. The seeds can be harvested in the fall.

How to gather it

You can pick the whole plant's leaves and stems when the plant is young in the spring. As it gets older, essentially before it flowers, you can harvest the younger growth off the top of the plant. To harvest seeds, cut the top of the plant off with the flower spike gently. Shake into a bag to gather the seeds and throw the top away.

How to cook with it

Edible: The young leaves and stems can be cooked like any other leafy green. The seeds can be boiled or ground into flour for baking.

WILD SPINACH COOKED GREENS
    Time: 18-20 minutes
    Serving Size: 4 servings
    Prep Time: 10 minutes
    Cook Time: 8-10 minutes
    Ingredients:

- 1 bunch lambs quarters (like spinach, when cooked it cooks down to what seems a 100th of its size, if you think you have enough you probably don't)
- 3 tbsps cooking oil, preferably olive oil or butter
- optional: foraged field garlic, garlic mustard, wild leeks, or wild onion for taste.

Instructions:

1. Wash the greens like any other lettuce and pick out any wilting or undesirable pieces.
2. Chop the plant up into edible size, this will also help the cooking process.
3. If you are not cooking the greens and are making a salad instead, it's a good idea to try the leaves pretty well after washing, especially if you are setting it aside for a time because fresh plants like this can become less crisp fairly quickly from the water.
4. In a pan, heat up the oil and throw in your spices, make sure garlic or onion is nice and soft.
5. Greens cook down fairly quickly, so it might be easier to add them a handful at a time. Stirring as you add more can prevent stress from trying to cook and stir a heaping pile.
6. Once all the greens are added, cook them all down for a few minutes, making sure they are stirred gently.

Author: I have a bias towards weeds as foragable food because I love a redemption arc. As we know, weeds are only named so because they are perceived to be so. That being said, it is not my favorite when I see it make an appearance in my own gardens. Weeds are always my go to when it comes to foraging because there is no guilt on how much I take. I know I am doing the surrounding plants a favor and that there is no short supply for the local wildlife as well.

63
LILAC

*T*he lilac is a name for the genus of plants named *Syringa*, but the common lilac is the most common, especially in the North east. The lilac is scientifically named *S. vulgaris*. While it is common in weedy, overgrown, or abandoned areas, it is not considered a weed. It can be a small tree or shrub.

IDENTIFICATION OF THE PLANT

The lilac is typically a bush, but the bush can range from anywhere between 3 to 20 feet tall. They often appear in small groups, becoming quite large. The bark is gray and smooth on the younger growth and becomes aged looking as it matures. The leaves are yellow green, basic oval point shape and up to five inches long. They are simple, growing in opposite pairs.

The flowers are in floret spikes that are fairly dense, sometimes more triangular shaped or more narrow. They can be anywhere between 3 inches to 10 inches long. The individual flowers are small with four petals. The shape is trumpet-like with a prominent narrow tube and the four petals open up flat. They are often a light purple, but can also be dark purple, pink, or white. They have a winged seed that is small.

Where to gather it

As noted above, it is common in weedy, overgrown, or abandoned areas like the sides of roads. Blooms in late spring.

How to gather it

Harvest the roses by bringing clippers strong enough to cut through the woody stems. Bring a basket that is big enough to carry the large florets. Be careful of bees and shake the florets gently beforehand to remove bugs. Since this is a flower that blooms in the spring, many bugs use it as a food source, you may want to leave them outside for an hour in the cool shade to let many of them escape before bringing them inside.

How to cook with it

Edible: The flowers are edible raw but are not prefered that way, although maybe lightly sprinkled in a salad. The flavor is very floral. This being said, there are many things you can do with lilac flowers. Most of the options are infusing the flowers in something else, like sugar, honey, or water.

Sugared Lilac Flowers

If you want to make more than this recipe, it's a good idea to have multiple small jars instead of one big one. It is really important to shake the jars everyday for at least a week or the water will clump the sugar.

Time: about 1 week

Prep Time: 20 minutes

Cook Time: 0 minutes

Ingredients:

- 1 cup white sugar
- ½ cup lilac flowers

Instructions:

1. Start to remove the small blooms from the stem and any hard part of the leaf. Note: this can take a while.
2. Submerge flowers in cool water to wash them. Let them dry completely before continuing.
3. In a clean jar, add a tbsp of sugar and the tbsp of lilac flowers.
4. Cover the flowers with a thin layer of sugar and repeat.
5. Put a lid on the jar.
6. Every day for a week, shake the jar to agitate the sugar and the blossoms.

Once the flowers are dry, you can strain the flowers or leave them in the sugar. The sugar will be infused with lilac and can be used in cooking.

Author: Lilacs are probably best known for their scent. Even wild rose bushes can smell too sweet for some people, but the lilac can be brought in the house for fragrance when it blooms as well.

64
LOTUS

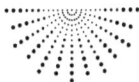

The he lotus flower is scientifically known as *Nelumbo lutea*. This is the only lotus native to America. They are not the same as a lily pad. The lily will sit in the water and the lotus will have a stem that pushes the plant above the waterline. Water lilies are poisonous.

IDENTIFICATION OF THE PLANT

The lotus is an aquatic plant that has two main parts above water. The leaf is the first part. They are round and can be over a foot in length. The stem pushes the above water, sometimes a foot out of the water. The pads are connected to the stem from the button in the center. This can create a cupping shape of the leaf, they are slightly wavy as well.

The flower sticks above water, typically higher than the leaf. The flowers seem disconnected completely from the pad. The flower is about seven inches wide and white to yellow, usually a pale yellow. Its flowers are a shallow cup shape, only curling up at the ends usually. The center is about an inch wide and tall, like a cylinder. It is surrounded by a thick ring of stamen.

Where to gather it

Grows in ponds and other slow or still water. The underwater stem gets about six feet tall, so it has to reach the bottom of water. Blooms late spring early summer. Roots are harvested in the fall. It prefers rich soil. Harvest seed after flower is done.

How to gather it

Harvesting all parts of this plant can be difficult since it is an aquatic plant. Finding plants that are located in shallow areas that can be waded through with rubber boots is best. The solid should be soft, so you might be able to pull the root up by the stem. Tug gently and see if it can be done without breaking the stem. Alternatively, use a small boat to reach plants in deeper water.

How to cook with it

Edible: Yes, all of it can be eaten.

The most popular part to eat is the root and large seed, but everything can be eaten. The leaves are preferred when they are young. The seed is eaten cooked or raw, and the root is cooked like a potato.

The greens are used like spinach. Seeds are eaten like nuts or starch.

ROASTED LOTUS ROOT
   Time: 45-50 minutes
   Serving Size: 4 servings
   Prep Time: 5 minutes
   Cook Time: 40-45 minutes
   Ingredients:

- 4 cups Lotus root

- ¼ cup olive oil
- 4 tbsps honey
- garlic, pepper, and salt for taste

Instructions:

1. Preheat the oven to 400 °F.
2. Clean and chop root into 1 inch pieces.
3. Lay on the tray evenly and sprinkle oil, honey, and spices evenly.
4. Roast for 30 minutes and flip it, while checking for tenderness.
5. Cook for another 10-15 minutes.

MAPLE

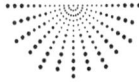

aple is a name for a genus of trees that go by the scientific name *Acer*. The maple is a national symbol of Canada. The sugar maple, *A. saccharum*, is most popular because of the maple syrup that it produces. It is commonly referred to as a sugarbush when there is an area of a forest that is being tapped for its sap.

IDENTIFICATION OF THE PLANT

The tree can grow up to 100 feet. It has a bark that is deeply grooved, rough textured, and dark gray.

Where to gather it

The sugar maple is a common tree, but it prefers areas that are not highly elevated. The trees can be tapped in the late winter to prepare for when the liquids start flowing come the warmer weather in fall. They grow in opposite pairs.

The leaves can range from a yellow green to dark and bright red to orange and yellow in the fall. The leaves are about eight inches long with five main lobes and are notched.

The seed is known for its spinning through the air. It has two connected seeds and a pair of papery wings.

How to gather it

The tree is drilled into with a shallow hole. A metal tap is inserted to direct flow and a bucket is hung to collect the liquid. Collect buckets regularly until there is enough. Seeds can be plucked off three while still green.

How to cook with it

Edible: The sap is made into maple syrup. The seeds can be boiled, then roasted.

MAPLE SYRUP

Serving Size: 4 servings

Ingredients:

- 3 gallons sap yields 12 cups

Instructions:

1. Boil outside only.
2. In a wide pan, add water and sap.
3. Add more sap as water dissolves.
4. When all has been added, bring to 4 degrees above boiling point.
5. Filter and add to jars.

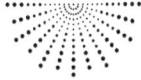

*T*he mayapple is a family (one above genus) of flowers scientifically called *Podophyllum*. They are also called mandrakes or ground lemons. The edible species is called just mandrake or may apple, and it is scientifically called *P. peltatum*.

IDENTIFICATION OF THE PLANT

The plant is only about a foot and a half tall. It usually grows in patches and can creep. The leaves are compound looking but are actually just deeply lobed. They attach to the plant from below in the middle, creating a circular fan. There are five lobes and a few large serrations, reaching up to 14 inches across. For the two leaves, the plant only slightly branches.

The flower is solitary in between the two branches. It is about two inches wide, cup shaped, and facing out, not up. The stamen in the center is short but prominent. The fruit looks like a miniature lemon, being only the size of a large grape. On the inside it is orange with four pairs of tear

248 | NORTHEAST FORAGING FROM YOUR BACKYARD H...

dropped hollow holes in a ring around the center that is softened when ripe.

Where to gather it

You can find this plant in wooded areas, meadows, or banks as long as it's rich and moist. Harvest ripe fruit in fall.

How to gather it

The plant should only be eaten when completely ripe. The fruit should be just about to fall to the ground.

How to cook with it

Edible: Only the ripe fruit can be eaten, otherwise it'll be poisonous. It can be eaten raw, but it is not for everyone. Can be made into jams and other preserves or used like lemon in juices.

Author: The root is used for treating parasites, among other things.

MELILOT, SWEET CLOVER

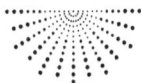

*S*weet clover is known scientifically as *Melilotus albus*, although other common names include honey clover.

This clover can be compared to other common clovers, the <u>red clover</u> and the <u>wood sorrel</u>. There is a white and yellow sweet clover, not to be confused with white clover. White clover more closely resembles the red clover in flower, the sweet white clover being a long spike, not a global foret.

Warning: If this plant is improperly dried, it will create a mold that can cause internal bleeding and **death.** Yellow sweet clover is often avoided.

IDENTIFICATION OF THE PLANT

This plant can grow up to 7 feet tall, with semi-erect, smooth stems, and sparse branches that are only 1-2 feet long. The leaves are typical of a clover, a compound of three, but unlike the other clovers in this book, the leaves are much more elongated and narrow, so it is not as noticeable as a clover.

The flower is a narrow spike at the top that is up to five inches long. The flowers are white, spaced out but not sparse, and often opened at the base, only budding at the top. The individual flowers are small and non-distinct, tubular shaped.

Where to gather it

This plant can be bunched in with other common weedy plants. It grows in abandoned areas.

How to gather it

Harvest shoots by cutting the young, tender plant before it matures.

How to cook with it

Edible: The flowers and young leaves are dried fresh for their vanilla flavor.

Author: This plant might be a great choice for people who are really stepping away from grocery stores and want to find good alternatives. In this case, vanilla. Since inpoper storage can be deadly, I do not recommend this plant be used unless you are experienced in foraging and can be confident in your skills to properly dry it.

68
MINT

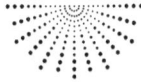

*M*int is a genus of plants that are known by the scientific name *Lamiaceae*. There are many common usable plants in this genus like peppermint, spearmint, catnip, and more. There are, however, members of this genus that are not edible like creeping charlie and more.

Mint is also invasive, if you are thinking about planting members of the mint family be sure to cut them back when you see that they are taking over your garden and the local environment.

Mint is probably best known for its cooling sensation and has many uses.

IDENTIFICATION OF THE PLANT

The mint family are able to be identified by their square stems. Peppermint, spearmint, and catnip all tend to stay under a foot tall, but can become bushy looking. All mint families have leaves that are opposite pair patterns. Peppermint tends to have darker leaves and a red stem. Spearmint tends to be yellow green and more herbaceous. Catnip is somewhere in between with a more bluish green color. Leaves on all of these are mighty wrinkled with scalloped edges. They are narrow, elongated heart shapes.

Catnip flowers with an elongated purple trumpet shaped flower that is on a small spiked floret. Spearmint is similar but the smaller spike is much more dense with much smaller individual flowers. The peppermint flower is almost identical to the spearmint flower.

If you are unsure whether a plant is mint, give it a smell. If it smells minty, give it a taste.

Where to gather it

The plant is invasive, but they like well drained soil that is well watered. They are common in abandoned areas.

How to gather it

Harvest the leaves any time of year that they are out. Since the plant is so invasive, you don't need to worry about giving it time to establish.

How to cook with it

Edible: Mint is a pretty diverse plant when it comes to cooking, being used in sweet and savory dishes. Most people use the fresh or dried leaves but the flowers are edible too. Mint is often used to settle stomachs and clear sinuses as well.

A list of ways to use mint:

- Mint tea
- Mint jelly (used on meats)
- Mint ice cream
- Mint candy
- Organic toothpaste
- Mint drinks
- Mint pea soup

- Aromatherapy
- Mint, caper, and lemon sauce
- Mint in salad
- Mint with chocolate
- And more

MUGWORT

*M*ugwort is a common name for a few plants in the genus *Artemisia*. Most people are referring to *A. vulgaris* when they refer to mugwort though, which is also called wild wormwood, St. John's plant, and more. This, of course, can be confusing since this plant can be confused with *A. absinthium*, which is commonly called wormwood. Wormwood has a silvery color and is more ornamental. Its name might also be confused with St. John's wort, which is again different.

IDENTIFICATION OF THE PLANT

This plant grows up to two and a half feet tall. The stem is red, hardy, and branching. The leaves are up to eight inches long. They are deeply lobed looking, like individual narrow leaves with lobes on a compound stem.

The flowers are narrow spike florets that are only a few inches in length made up of small white to pinkish purple flowers. Their petals look stringy.

Where to gather it

The plant can be invasive and is found in other weedy areas. It flowers in July.

How to gather it

Harvest shoots by cutting the young, tender plant before the plant flowers.

How to cook with it

Edible: The leaves are eaten raw or cooked, sometimes used as tea. It is more commonly used as a bitter spice. It was traditionally used for flavoring in beer.

MULBERRY

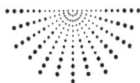

The mulberry refers most commonly to three mulberries: the white mulberry, scientifically known as *Morus alba* and used for the cultivation of silkworms, which are used to create silk, the red mulberry, known as *Morus ruba*, and the black as *Morus nigra*. They are not sold in stores because of the green stem they are attached to.

IDENTIFICATION OF THE PLANT

The mulberry can take form as a small tree or shrub, but can grow up to 60 feet on occasion. On the white and red mulberry, the young twigs are green and smooth, the other plant is brown and woody. The leaves on the young shoots are up to 12 inches long and deeply lobes into about three. The lobes are not always symmetrical. On older growth and the black mulberries, the leaves are only about five inches long and simple. Both have small serrated edges.

The flower is not noticeable, being small, sparse string-like petals on a green base. The berry of the mulberry looks similar to a blackberry, but the tiny sacs are less uniform and plump. They can be around the same shape or much more elongated to an inch and a half. The berries start off as white or green (wild blueberries typically have a variation of the three), then to red (red mulberries might be red or dark), and then parts of it get dark purple to black (black mulberries always black).

The black mulberry has hair on the bottom of its leaves.

Where to gather it

The white mulberry is invasive and found in thickets and other overgrown areas. The red and black mulberry prefers very moist places like flood lands and the like. All berries are ripe in the early summer.

How to gather it

With a ladder, you can pick the berries or lay a tarp and shake the bush to release the berries.

How to cook with it

Edible: Yes, the berries are eaten.

Do not eat unripe fruit or shoots. It can cause hallucinations. The red and black berries are preferred because they are sweeter. They are used similarly to other berries, see brambles. Spring leaves can be boiled.

Author: Although mulberries are similar to other berries, they don't grow on brambles and have slightly different nutritional information. Mulberries stain, so avoid white clothing and table clothes.

MUSTARD

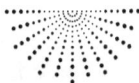

ild mustard is also known as field mustard or scientifically as *Sinapis arvensis*. Field mustard is scientifically known as *Brassica rapa*. Despite the scientific name, they are related. Other species of mustard such as *B. nigra* are also edible. This is also the same family that regular store mustard is made from as well. Be careful when handling this plant as it can cause rash and irritation. It is best to use gloves when handling this plant raw.

IDENTIFICATION OF THE PLANT

Wild mustard and other mustard varieties grow up to two feet high and are erect. Its stem is green with white hairs. As it gets older and taller it branches more, looking strange. The young leaves are long, oval, slightly tapered to the end, and slightly wrinkled in texture. The edges are slightly jagged or scalloped. Up the plant the leaves get smaller, up to one inch, and smooth.

The mustard plant has a yellow flower with four petals that looks like a buttercup but is part of a small round cluster of other flowers. The fruit is hard to notice, being an inch or so long and slightly wider than the stem, kind of like a hairy bean, with small black seeds inside.

Where to gather it

Wild mustard is invasive in many places. It grows everywhere in the world. It flowers from early summer to fall.

How to gather it

The leaves can be harvested by cutting them with gloves on. The seed pods can be gathered in the summer through to the fall when they appear.

How to cook with it

Edible: Yes, seeds, roots, and leaves are edible.

The seeds of wild mustard can be dried or fresh and used as a spice. The leaves can be cooked into leafy greens, like spinach but bolder in flavor.

MUSTARD

   Time: 24 hours
   Serving Size: 4 servings
   Prep Time: 24 hours
   Cook Time: 0 minutes
   Ingredients:

- ½ cup mustard seed
- ½ cup water
- ¼ cup white vinegar
- pinch of salt

Instructions:

1. Remove seeds from seed pod.
2. In a mortar and pestle, gently crush.
3. In a jar, put the crush seed.

4. The next step is a chemical reaction when the water is added. The vinegar stops this reaction. The chemical reaction is instant, so add the vinegar right away for stronger flavor or wait for a more mild flavor.
5. Add salt and whatever other spices you like.
6. Let sit to thicken overnight.

Author: Did you know that mustard helps digest meat? This is the most amazing mustard recipe I've ever found! Try it once and you'll stick to it. It's very easy to preserve for longer periods of time in the fridge.

NASTURTIUM

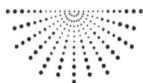

*N*asturtium is a genus of flower scientifically named *Tropaeolum*.

IDENTIFICATION OF THE PLANT

Nasturtium is a vining herbaceous plant that likes to climb. Its leaves are circular on long stems, slightly scalloped, and a pale blue-green color. The leaves resemble lily pads. The flowers have five petals that form a shallow trumpet shaped and delicate petals that range from yellow to red, sometimes multiple.

Where to gather it

While nasturtium does not grow wild in the northern part of the US, it is a common flower found in flower gardens and is a surprisingly great addition to your plate. It can be planted in poor quality soil and partial shade.

How to gather it

Once the plant has established itself in early summer, the leaves and flowers can be picked. Picking the flowers regularly encourages more growth.

How to cook with it

Edible: The leaves and flower have a strong peppery flavor that is unexpected with a flower. They are used fresh when used to cook with. Think of them like a garnish you can add at the last minute so as to not overcook it. Seeds can be pickled and used instead of capers.

See <u>Yarrow</u> for recipe.

73

NETTLES

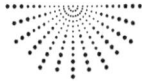

*N*ettle is a genus of flower called *Urtica*. Stinging nettle or *U. dioica* is a plant that can cause **severe skin reaction**, but when handled carefully and cooked right is actually edible. Its relatives are wood, edward, and slender nettle, or otherwise known as *Laportea canadensis*, *U. gracilis*, and *U. urens*, that look like the stinging nettle, but they are also edible.

Also see henbit and deadnettles for more on nettles.

IDENTIFICATION OF THE PLANT

Stinging nettle can get up to four feet tall and erect. Its stem is lined with hairs, those are the nettles. Leaves are covered in the hair as well. They are opposite in pattern. The shape of the leaves are narrow and triangular with jagged edges. They also have a wrinkled texture.

The flowers are semi drooping spikes about four inches long. There are many of them. The individual flowers are whitish green and tiny. The shape is not distinguishable.

Where to gather it

The plant can be found in many areas because it is invasive. However, it prefers places with fertile soil. Harvest it in the spring.

How to gather it

Harvest young leaves by wearing gloves and other protective gear. Be careful not to let it touch your skin. Cut the leaves and follow the steps below to make it edible.

How to cook with it

Edible: Yes, the cooked leaves are edible.

STINGING NETTLE PREP

Time: 10 minutes
Prep Time: 2 minutes
Cook Time: 8 minutes
Ingredients:

- 4 cups stinging nettle
- 1 tbsp vinegar
- 1 tbsp butter
- pinch of salt

Instructions:

1. With gloves on, wash the nettle.
2. In a pot, add water and nettles and bring to a boil.
3. The boiling destroys the stinging chemical.
4. Remove leaves.
5. Add butter, salt, and vinegar for taste or use the cooked nettle in a different recipe.
6. Water can also be used as a drink or soup.

Author: Some may wonder why go to the trouble of risking a rash with this plant, and the answer is that it is high in nutrients. It also is used in herbal medicine to treat many things, mainly urinary tract infections.

74

OAK

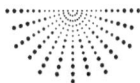

The oak is a genus of trees scientifically called *Quercus*. The most edible oak in the Northeast include white oak (*Q. alba*), chestnut oak (*Q. prinus*), and Northern red oak (*Q. rubra*).

IDENTIFICATION OF THE PLANT

The white oak gets up to 100 feet tall. Its bark is furrowed and a light gray. The leaves are alternate, about nine inches long, and lobed up to 10 times. The chestnut oak is about 70 feet tall with darker red brown. The leaves are very similar with a more prominent middle point. The red oak is about 65 feet tall, leaves similar.

The fruit is the **acorn**, which is pretty similar for each tree and fairly recognizable to most. It is a cherry sized nut with a textured top that has a stem like a barret.

Where to gather it

The oak is a fairly common tree. The white oak is found in sandy forests, chestnut oak is found in moist forests, and the northern red oak is found in hilly areas. Gather acorns in autumn.

How to gather it

Harvest the nuts off the ground when they fall from the tree. They should be brown and intact. Make sure there are no blemishes or holes.

How to cook with it

Edible: Some acorns can be eaten raw. The nut of the Northern red oak must be cooked, since it can be poisonous. They can be used as a nut or ground into flour.

NORTHERN RED OAK ACORNS

The nut can be ground into flour and used best in baking heavy foods like muffins or dense bread, they can be candied or eaten as is. When roasted, it can be ground and used to make a coffee blend.

Time: 17 minutes
Prep Time: 2 minutes
Cook Time: 15 minutes
Instructions:

1. In a pot, bring water and acorns to a boil.
2. The acorns can be whole or chopped, but remove the caps.
3. Boil until the water is brown, then dump the water and refill.
4. Bring new water to a boil, repeat until water remains clear.
5. Once done, dry acorns in the oven or in the sun until completely dried.

75

OSTRICH FERN/ FIDDLEHEADS

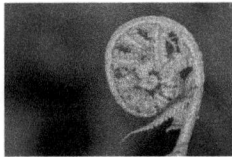

*T*he ostrich fern, also known as *Matteuccia struthiopteris*, is the only species in the genus. Other ferns are used as fiddleheads as well, but not all of them. Ostrich fern as well as bracken (*Pteridium*) are found in the Northeast and are edible.

IDENTIFICATION OF THE PLANT

Fiddleheads are the young stalks of ferns. They are recognizable by the short green stem that curls up into a spiral at the stop which will eventually unravel into a fern. An identifying characteristic is a brown paper that protects the young fern that eventually falls away.

If you are identifying a grown fern for reference for next year, they have opposite pairs with no spacing. The leaves are scalloped and 2-3 inches each, narrow, and end at a point. The leaves come out of stems that come from a centralized spot in the ground, this is unlike some ferns that have a stem and branch. They can create a feathery bushy look. They get their name because they resemble ostrick tails.

Where to gather it

The plant can be found in damp soil areas, especially around water. They prefer fertile soil. They are harvested in the spring.

How to gather it

Harvest the plants that are up to eight inches by cutting them at the base.

How to cook with it

Edible: Yes, the young curled plant can be eaten raw, but people usually prefer them cooked.

SAUTEED FIDDLEHEADS

Time: 22 minutes

Serving Size: 4 servings

Prep Time: 2 minutes

Cook Time: 20 minutes

Ingredients:

- 3 cups fiddleheads
- 3 tbsps butter or olive oil
- 1 pinch of salt, garlic, and pepper for taste

Instructions:

1. Bring water and salt to a boil and add fiddleheads.
2. Cook for about 10 minutes or until tender.
3. In a pan, heat butter and spices to a sizzling heat, then add drained fiddleheads.
4. Cook until there is a bit of browning.

Author: Fiddleheads are considered a delicacy, and they are often fairly hard to get. The season is short, but timing it right is worth trying. Once I picked a plant that looked almost the same as an ostrich fern, and the first bite I took was seriously the most disgusting thing ever.

76

DAISIES

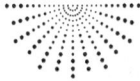

*D*aisy is a common name for a type of flower. As a family of flowers, one branch above a genus, it is a very large group. Under the family, there are multiple genus, and under the genus, multiple species. Some of the popular plants in this group are feverfew (*Tanacetum parthenium*), chamomile (*Matricaria recutita*), and the oxeye daisy. The flowers listed are just some of the flowers in this family that look like the common daisy that we know. There are many more varieties in this family that look completely different. Because this is a big family, it's important to identify closely.

In this chapter, we are going to focus on the three plants listed above.

IDENTIFICATION OF THE PLANT

All three of these flowers are well known for their golden yellow center and fan of white petals all the way around the edge. The oxeye daisy bloom is quite a bit larger than the other two. The petals on feverfew tend to be short and rounder than chamomiles as well.

The body of the oxeye daisy is spindly with only a few long stems. The other two are more like small shrubs, but feverfew tends to be a little harder and erect, while chamomile is more herbaceous and can end up almost creeping across the ground.

The leaves of chamomile are thin, like feathers or dill. Feverfew is about an inch long and deeply lobed. Oxeye daisies have small narrow leaves on the stem, but at the base have long three or four inch leaves that are rounded at the end, slightly reverse scalloped on the edges.

Where to gather it

All of the flowers are planted as flowers in the garden but are also seen as a weed in some places. They can grow in overgrown or weedy areas. They like well-drained soil. The oxeye daisy flowers from late spring to fall. Feverfew flowers a little later, in early summer, but goes a little longer into fall. Chamomile blooms in spring and summer but only if it has full sun.

Feverfew smells citrusy.

How to gather it

The blooms can be cut off throughout the season, this might even encourage more growth.

How to cook with it

Edible: The oxeye daisy unopened blooms are edible and usually picked. The head is edible but not preferred. The leaves can be eaten raw or cooked. Root can be eaten raw.

The feverfew flower and leaves are used fresh in cooking as a citrusy/ bitter spice or dried as tea.

The chamomile is used as a tea. Leaves can be eaten but taste bitter.

DAISY SALAD
Time: 10 minutes

Serving Size: 3 servings
Prep Time: 10 minutes
Cook Time: 0 minutes
Ingredients:

- 1 oxeye daisy root
- ¼ cup chamomile heads
- 2 cups oxeye daisy flower
- 2 tbsps chamomile infused oil
- 4 tbsps apple cider vinegar
- ½ tbsp honey
- 4 tbsps raspberries

Instructions:

1. Clean all of the plants well.
2. Chop root and leaves to bite size.
3. In a bowl, add oil, vinegar, honey, and 2 tbsps raspberries.
4. Mix together, crushing the berries as you go.
5. In a salad bowl, add the root, flowers, and leaves, then toss.
6. Drizzle dressing on top.
7. Serve as a side salad.

Author: The daisy family is similar to the rose, in that they are known for their beauty and useful in other ways too. Chamomile and feverfew are used for health purposes like calming, sleeping, headaches, infection, and so on. It is a great idea to get familiar with this plant and find ways to implement it in your life.

PARSNIP

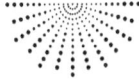

The wild parsnip is a species scientifically called *Pastinaca sativa*.

Warning: Wild parsnip can cause severe skin irritation, severe blistering, and burns. Use with caution.

IDENTIFICATION OF THE PLANT

The wild parsnip grows about two feet tall. The stem is herbaceous but thick. The branches come out in opposite rows. Many branches end in florets. The lower leaves have three lobes. They are compound, up to a foot long. The leaves on the floret are opposite rows of pairs. The leaves near the base are longer and have more points than the leaves at the top of the leaflet.

The flower is a round and flat topped floret. It consists of a few branches that end in a hollow circle of tiny yellow flowers. The circle doesn't touch, making it look delicate looking. The flowers are about eight inches across at the largest. Seeds replace flowers.

The root is long and carrot-like but white.

Where to gather it

The plant is a common weed that can be found in abandoned spaces. As such, it is very common in fields and can be harvested all year round, although it's best in the fall or winter.

How to gather it

Harvest the plant wearing complete protective gear. Pull out roots from the ground. Put the plant foliage inside a garbage bag and leave it in the sun for a week to dry it out. Be sure to dispose of it properly.

How to cook with it

Edible: The roots are considered very tasty even though the foliage can be so damaging. The recipe below is how to prepare and cook this member of the carrot family.

CARROT AND PARSNIP PUREE

Time: 30 minutes
Serving Size: 4 servings
Prep Time: 5 minutes
Cook Time: 25 minutes
Ingredients:

- 2 cups parsnip
- 2 cups carrot
- 2 tbsps of butter or olive oil
- a pinch of salt, pepper, dill, nutmeg, or cinnamon.

Instructions:

1. Bring water and salt to boil.
2. Wash, peel, and chop the roots.
3. Add to the pot and cook until they are soft (less than 20 minutes).

4. In a blender, add the strained roots and everything else.
5. Blend until smooth.
6. Serve like mashed potatoes.

Author: Roasting is super simple and brings out all the deeper flavors of the veggies you're roasting. But as delicious as roasted carrots or broccoli can be, roasted parsnips really take the cake! Honestly. These yummy root veggies are naturally sweet and earthy, and they really show off their flavors after roasting.

7 8

PAWPAW

*T*he pawpaw plant is scientifically called the *Asimina triloba.*

IDENTIFICATION OF THE PLANT

The pawpaw plant is a shrub or small tree that grows up to 40 feet. The bark is smooth and dark. The leaves are up to 12 inches long. They alternate on reddish stems. They have parallel veining. The leaves are narrow at the base, begin to widen up the plant, and then end in a point.

The flower is dark red/burgundy. The shape is a trumpet/rosette. There are three smaller pointed petals in the center and, in their alternating spaces behind them, are larger, more rounded petals. The petals also have deep vertical veining causing seams. The center is large, yellow with a slightly raised point. They can slightly droop from the branches. The fruits are peanut-shaped and green. They grow to about four inches tall.

Where to gather it

The plant is found in patches of well-draining fertile soil.

A good place to look is in valleys. The fruit is ready to pick in the fall.

How to gather it

Harvest the fruit when they are soft. Give them a twist off the plant. You may need a ladder to reach. When the fruit is ready to pick, they turn a black color. You can also pick when green and then store.

How to cook with it

Edible: Yes, the fruit can be eaten raw or used in cooking in both sweet and savory dishes. Can be used like squash or bananas. Avoid seeds and skin.

Pawpaw bread

Time: 1 hour, 5 minutes
Serving Size: 8 servings
Prep Time: 5 minutes
Cook Time: 1 hour
Ingredients:

- 1 ½ cups pawpaw
- 1 cup sugar
- ⅓ cup butter
- 2 eggs
- 1 ½ cups flour
- 1 tsp baking soda
- pinch of salt
- ⅓ cup water

Instructions:

1. Preheat the oven to 350 °F.
2. In a bowl, combine sugar and butter. When smooth, beat in eggs.

3. Add the rest of the ingredients and mix.
4. Pour into a bread pan.
5. Bake for 1 hour.
6. Check to see if cooked with a toothpick, add more time if needed.
7. Let cool and slice.

PEACH

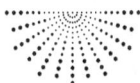

*T*he peach is a common fruit that is scientifically named *Prunus persica*. This means that the peach is actually just one species, but it doesn't mean that there are no further variations of the plant, just that they are all related under the species. For example, nectarines are actually a variant of the peach. These variations are all thought to be probably edible, but usually not as pleasant. If you find the peaches from your tree are having these issues, it might be a variant like the ornamental peach tree that is, as the name implies, used for ornamental purposes. If your peach tastes good and juicy, then you are good to go.

IDENTIFICATION OF THE PLANT

The peach tree is about 20 feet tall, but the branches stretch out wide. The bark is dark and smooth, sometimes slightly peeling. The branches often start lower on the trunk and end up thick like two trunks. The leaves reach up to six inches long. They are narrow at the base, wide, and have a prominent, elongated point. They taper in a bit between the base and widest width, giving it a slightly peanut shape.

The flower of the peach tree grows in small clusters of 1-4 out of the branch. They put on a great show in the spring. The flowers are pink. They are about 1-2 inches big. Some of them have a cup shape with a single row of petals, while some are more rosette-like with multiple rows, but still open slightly in the middle. The fruit is five inches wide. The skin of a peach is thin and soft. The color of the peach is a yellowy orange, sometimes with darker purple tone or yellow tone patches. The skin often has a velvety feel but also can be smoother. The flesh is orangey-yellow, fleshy, juicy, and sometimes a bit stringy. The wrinkly pit is in the middle.

Where to gather it

The tree is often planted, although it might be found spreading in areas with warmer climates. They like sandy soil and full sun. The fruit is ready from early summer to late summer.

How to gather it

Harvest the peaches by twisting them off of the tree. Test to see if the peaches are ripe by feeling their softness or giving them a taste test. If picked too soon, they might not get sweet. If picked too late, they will go bad very quickly. If you are picking blooms, don't take too many as these will become the fruit. Leaves can be picked when the plant is fruiting.

How to cook with it

Edible: The fruits can be eaten raw or cooked into many dishes. They can be preserved in syrup. The blooms can also be eaten, usually only used for decoration because they are mild. The leaves can be eaten but must be boiled. They are used in sweet dishes or tea.

. . .

Peach Cobbler
  Time: 1 hour
  Prep Time: 15 minutes
  Cook Time: 45 minutes
  Ingredients:

- 4 cups (about 7) peach
- ½ cup butter
- 1 1/4 cup all-purpose flour
- ½ cups sugar (brown recommended)
- 1 tbsp baking powder
- pinch of salt
- 1/4 cup buttermilk
- 1-2 tbsps lemon juice
- 1 tsp Vanilla extract
- ground cinnamon or nutmeg
- ice cream

Instructions:

1. Clean and chop peaches into ½ inch pieces.
2. Combine flour, half the sugar, baking powder, and salt.
3. Mix in milk and pour the batter in a buttered pan.
4. In a clean bowl, add sugar, peaches, lemon, spices, and extract.
5. In a pot, bring to a boil.
6. Pour over batter. Don't stir.
7. Bake at 350 °F for about 40 minutes.
8. Serve hot or cold with ice cream.

PEAR VS PRICKLY PEAR

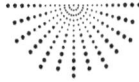

*P*ear is a common name for fruit that grows from trees, the prickly pear is a cactus. It might be shocking to consider the possibility that there are cacti in the northeast, but the Eastern prickly pear grows native to New England. The eastern prickly pear is scientifically called *Opuntia humifusa* and the pear tree is part of the genus *Pyrus*.

Appearance is not really an issue when it comes to the difference between the prickly pear and the pear, but the name can be confusing.

There are pear trees that are considered poisonous and invasive; take, for example, the species *P. calleryana*.

IDENTIFICATION OF THE PLANT

Pears are small trees that grow up to 60 feet but often can be shorter. The leaves are 1-2 inches long, basic oval point shaped, smooth, darker green, and thick in texture.

The flowers are up to two inches wide and star shaped. The petals are round and spaced. The stamens are prominent and tipped with darker ends. The fruit is pear shaped (obviously) but can be more round, with short stems. They are usually yellow to pale green in color.

The **prickly pear** is a cactus that is short, about 1-2 feet tall. It is composed of sections that are oval-shaped disks that stack vertically onto each other. They are green with widely distributed thorns.

They bloom in spring. The flowers are yellow and cup shaped that are 4-5 inches wide. The fruit grows in 4-5 inch egg shaped red fruit that grows out of the top of the disks. The fruit is yellow to red.

Where to gather it

Harvesting the pears is done in late summer to fall. Prickly pears are ready late summer, but they can be harvested all year.

How to gather it

Harvest pears by twisting them off the tree when the pear is ripe. You can tell they are ripe and sweet when they are turning slightly yellow. You can pick them before ripening for longer storage, but they won't be as sweet.

Prickly pears are cacti and so you need to be careful and use gloves. Only take ⅓ of a prickly pear. The new growth grows in oval shapes on the main plant. Harvest mid-morning by twisting them off the main plant.

How to cook with it

Edible: Pears can be eaten raw and cooked in sweet and savory dishes. It can be dried and preserved as well.

Eastern prickly pear can be peeled and eaten fresh. They can be made into sweet syrups and jams, sometimes used for flavoring meats as well. Peel with gloves on or sear the skin off.

PRICKLY PEAR LEMONADE
      Time: 30 minutes

Prep Time: 5 minutes
Cook Time: 25 minutes
Ingredients:

- 1/2 cup prickly pear
- 1 cup sugar
- 1 cup water for syrup
- 1 cup cold water

Instructions:

1. Skin, clean, and chop the prickly pears.
2. Bring water and sugar to a boil.
3. In a blender, add prickly pear and puree.
4. Boil the simple syrup until it's reduced by half.
5. Take off heat and let cool.
6. Add to puree and mix.
7. In a glass, add cold water and mix in prickly pear syrup. Start with 1 tsp and add to desired strength.
8. Optional: add lemon.

Pear Meat Sauce
Time: 25-35 minutes
Prep Time: 5 minutes
Cook Time: 20-30 minutes
Ingredients:

- 2 tbsps butter
- 1/3 cup onion
- 2 tbsps ginger
- ½ pear
- 1 1/2 cups broth
- 3 tbsps vinegar
- 2 tbsps honey

- pinch of rosemary and ground black pepper

Instructions:

1. In a pan, add butter and onion and soften.
2. Add the pears until softened.
3. Add broth, vinegar, rosemary, and honey/sugar.
4. Reduce until thickened. If it takes too long, add a pinch of cornstarch.

PEPPERGRASS

$\cdots\cdots\!\!\ast\!\!\cdots\cdots$

*P*eppergrass, also known as virginia pepperweed, is scientifically known as *Lepidium virginicum*. It is invasive.

IDENTIFICATION OF THE PLANT

When it is young, it is a tuft of leaves growing out of a centralized point in the ground and folding outwards from it. It grows about a foot tall and looks bushy sometimes with multiple stems that are more densely packed. The leaves are about an inch long but are very narrow. The edges are serrated but get smooth when higher on the stem. In general, it is a weedy looking plant.

The feathers are a small florret on the top of the stem. They are spike shaped, but rounded out on top. It's more dense at the top and spaces out after an inch or so. The flowers are small and white with four petals in a star shaped pattern. It turns into a flat green seed.

Where to gather it

The plant can be found in weedy areas like places that have been abandoned. It prefers dry sunny areas.

How to gather it

Harvest leaves when they are young in the spring or early summer. The seeds can be removed from the stalk when they are green.

How to cook with it

Edible: The leaves can be eaten raw or cooked like spinach. The seeds can be dried or fresh and used for their peppery taste.

PINEAPPLE WEED

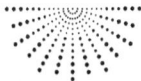

*P*ineapple weed might also be called wild chamomile or *Matricaria discoidea*. The plant is used medicinally to aid in digestion and infection, among other things.

The plant looks similar to chamomile and even smells similar. While they are different plants, they are part of the same genus and used similarly. The petals are almost completely unnoticeable; it might be not recognized as part of the <u>daisy</u> family.

IDENTIFICATION OF THE PLANT

This plant can grow about half a foot tall. The plant grows in bushy sections. The stems are yellow-green, smooth, and herbaceous. The leaves grow in opposite pairs. They grow up to four inches at the base and get smaller up the plant. The leaves are dark green and deeply lobed into thin points.

The flowers are at the end of a branch with a cupping

bracket. The petals are white but are so short that they are sometimes completely unnoticeable. The head of the plant is where it gets its name. It is a pointed dome shape that is yellow with tiny sections that make it look like a pineapple skin. It can get about half an inch tall.

Where to gather it

The plant can be found in woodland areas especially near banks, but they also like well drained soil. Gather shoots in the spring. Berries are ripe very late in the season, even into the later parts of fall.

How to gather it

Harvest shoots by cutting the young, tender plant before it matures.

How to cook with it

Edible: The greens can be eaten as leafy greens, cooked, or raw. The flower and the rest of the plant is used to make tea. The tea can be used similarly to chamomile tea for calming and to aid with sleep. Since this plant can aid with sleeping, it is best used at night, not as a morning pick me up.

DRYING FLOWERS

Instructions:

1. After you have picked the plants, clean them thoroughly.
2. Dry then completely with a towel.
3. It is best to avoid drying in the oven at the risk of cooking them and reducing their properties:
4. Option one is hanging and drying them in a bundle.
5. Option two is cutting the heads off and putting them in a mesh or fabric bag and hanging that in a place to dry.

6. Option three is laying them out evenly on a tray in the sun in the summer until dry.

83
PLANTAIN

*P*lantain, also known as broadleaf plantain or scientifically as *Plantago major*, should not be confused with the large banana-like fruit. Plantain is a common weed that grows in most gardens.

IDENTIFICATION OF THE PLANT

The plant is typically very small, only reaching a few inches tall; it blends into most grass lawns. The plant consists of a cluster of leaves that form a circle around a centralized spot on the ground. The leaves are round, sometimes more heart-shaped, sometimes leaning more narrow and pointed. The leaves can get fairly long, being about 12 inches. The leaf is textured with a wrinkle, sometimes more or less prominent. The flower is a stalk that stands erect a few inches. The floret is narrow, only slightly thicker than the stem. The individual flowers are indistinguishable. They are green and go a rusty brown.

Because it is part of many lawns, the constant mowing might keep the plant small. When young, its leaves are more yellow green, watery, and less textured. More noticeable on the young leaves are multiple faint parallel margins.

Where to gather it

The plant can be found almost everywhere, more specifically in places with grass. It likes well watered areas. They do really well in spring but can also be found through summer.

How to gather it

Harvest shoots by cutting the young, tender plant before it matures.

How to cook with it

Edible: The young leaves and seeds can be eaten raw. The older leaves should be cooked like spinach.

Author: The leaves and seeds can be dried and used for medicinal uses. The plantain is great to use on stings and other skin irritants.

84
POKEWEED

*P*okeweed is also called inkberry among other common names. It is scientifically known as *Phytolacca americana*. It can look like a number of plants from the *Fallopia* genus. However, they can be differentiated by the fact that pokeberries have berries, but since they are poisonous and the plant is harvested in the summer, this can be an issue. Pokeweeds leaves tend to be more narrow and its stems are paler. Looking at the previous year's growth, the pokeweed tends to fall, whiten, and decay where the buckweeds tend to stay standing. When they are being harvested, this can be important when distinguishing them from one another.

The pokeweed has been used in medicinal medicine, but it has not been proven to work.

IDENTIFICATION OF THE PLANT

The pokeweed can grow up to eight feet tall. It is a herbaceous plant with elongated oval-point leaves that alternate.

The stem is a dark pinky red or green. It can look tree-like because of the thick stem and the way it branches at the top.

The flower grows on spike clusters about five inches long. The stem is white, and the flowers are on steam that stick out straight from the main stem. They are small and star shaped with a round, raised, green center.

The berries grow on long, straight, drooping clusters about a foot long. They are narrow with pink stems and brackets. The fruits are green that ripen to a red, then black. The cluster branches the berries straight out on the stems. As such, they don't droop like grapes. The berry dimples in the middle of the bottom and seems to run from it to the bracket top. They are about the size of blueberries and slightly more flat.

The young plant looks very different at the end of the season. When young, it looks more herbaceous and green. When mature, it is hardier and pink.

Where to gather it

The plant can be found in open weedy areas. Common in abandoned areas or edges of forests.

How to gather it

Harvest shoots by cutting the young, tender plant before it matures. It is not recommended to touch it with bare hands.

How to cook with it

Edible: Yes, the young leaves and stems are edible, but the mature plant, roots, leaves, stem, and berries are poisonous. They are a traditional food for the south.

Poke Sallet

Time: 1 hour, 10 minutes

Serving Size: 4 servings

Prep Time: 5 minutes

Cook Time: 1 hour, 5 minutes

Ingredients:

- 4 cups pokeweed leaves
- 1 onion
- 2 tsps garlic
- 1 cup cooked rice

Instructions:

1. Wear gloves, then clean and chop pokeweed.
2. In a pot, bring to a boil, boil for 5 minutes, then drain the water.
3. Fill the pot up and boil for 30 minutes.
4. If it's not clear, repeat one more time.
5. Strain.
6. In a pan, add oil, onion, garlic, and other spices like ginger, pepper, and salt.
7. Add rice and fry it so it starts to brown.
8. Add pokeweed.
9. Serve with soy sauce.

85

PURPLE DEAD NETTLE

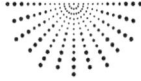

*P*urple dead nettle is called red dead nettle, also known as *Lamium purpureum*. It has a look-alike, the henbit, which is also used for eating. As discussed in that chapter, the name dead nettle can be alarming. Nettles are actually the dangerous part, but the 'dead' descriptor means that the nettle is dead, not deadly. As such, it will not sting you.

Dead nettles are safe to handle, but they can look similar to the nettles. It is important to identify with complete confidence that the plant you are working with is the deadnettle before you touch it. The worst thing to not know is that they often grow together. The stinging nettle leaves tend to be a rounder and a bluer green. The large serrations on the edge are longer.

IDENTIFICATION OF THE PLANT

The purple dead nettle grows up to half a foot tall. The leaves and stem have fine hairs (the dead nettles). The leaves turn green at the bottom and get to a purple on the top. The green leaves are wrinkled with a red tinge on the edges. They are teardrop shaped with short serrations on the edges.

Flowers are a magenta pink. They grow right out of the stem. They are long, tubular, and open at the end to a non-symmetrical formation. Two petals open out away from the plant, one petal on the other side is pointed up and over, like a hood. They tend to be less than the ones on the henbit, and are also smaller and more narrow.

Where to gather it

The plant can be found in many areas because it is invasive. However, it prefers places with fertile soil. Harvest it in the spring.

How to gather it

Harvest shoots by cutting the young, tender plant before it matures. Be aware of how early the plant's flowers need to be harvested very early in the year.

How to cook with it

Edible: Yes, the young plant is edible.

The leaves can be eaten raw or cooked, similar to stinging nettle.

Author: The first time I went out with my kids to forage for some purple dead nettle they thought I'd trick them into getting stung because they know what happens when they touch stinging nettles which look really similar to these.

This plant is used topically for healing wounds.

QUICKWEED

*Q*uickweed is a herbaceous plant that is also called potato weed or *Galinsoga parviflora*. This plant is called many different names across the world.

The plant is applied to the skin to stop the stinging nettle rash.

IDENTIFICATION OF THE PLANT

It grows up to half a foot tall. It is weedy looking and almost looks like it is crawling. Its stem is semi erect. The leaves are opposite pairs, 1-2 inches long, basic pointed oval shape on bigger leaves and more narrow on smaller ones. The stem and underside of the leaves are covered in long white hairs.

It branches slightly, and they are topped with a few flowers. The flowers are similar to a daisy, but are much smaller, about ½ inch wide. The petals are small, short, and round. There are only five petals and they don't touch. The center is yellow, slightly domed, and has a honeycomb pattern.

Where to gather it

The plant grows in weedy areas like overgrown places or meadows. The plant blooms all summer.

How to gather it

Harvesting plants by cutting the young leaves, flowers, and stems.

How to cook with it

Edible: The leaves can be used fresh or dried as a spice. It can also be used as a leafy green, including the stem and flower.

Ajiaco

Time: 43 minutes

Serving Size: 4 servings

Prep Time: 8 minutes

Cook Time: 35 minutes

Ingredients:

- 1 tbsp oil
- 1 chicken breast
- 1 onions
- 2 garlic cloves
- 4 cups broth
- pinch of salt and pepper
- 1 cup cilantro
- 1 green onion
- 1 tbsp Quickweed
- 1 pound mixed potatoes
- 1 ear of corn
- 1 cup cooked white rice
- ¼ cup heavy cream
- ¼ cup pickled juniper

Instructions:

1. Put oil, onion, garlic, and chicken in a pan and brown.
2. Add broth, green onion, spices, and quickweed, then bring to a boil.
3. Reduce to a simmer and cook for 20 minutes.
4. Take out chicken when cooked.
5. Add chopped potatoes and corn to broth.
6. Shred chicken.
7. When potatoes are cooked, add the chicken back in.
8. Add rice, cream, and juniper to serving bowls.
9. Pour soup on top.

Author: When I was helping out at a mountain hut, we'd serve this salad called "foraged greens." Only 1 in 100 people wouldn't like it because of the furry leaves. Everyone else was always super stoked about it.

RAMPS/WILD LEEK

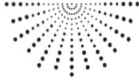

*R*amps or wild leeks are known scientifically as *Allium tricoccum*. They are under consideration in Canada and some states in the US because they are a threatened species.

IDENTIFICATION OF THE PLANT

The leeks usually grow in patches. They appear above grown as a crown base (leaves that go from a centralized spot in the ground). The leaves are long, rounded blades. They are a little less than a foot long and an inch or so wide. When pulled from the ground, you will notice a red stem that ends in a white bulb. They also grow a stem that is about a foot tall with a floret at the top. The floret is a sparse global shape. The individual flowers are small, white, and cup shaped.

Noticeably smells like onion/garlic.

Where to gather it

Ramps like wet areas, like lowlands, more specifically wooded areas that flood in the spring. Available in the very early spring.

How to gather it

Gather the leaves in your hand and grasp the base at the ground firmly. Carefully pull the leek out of the ground, keeping the bulb intact.

How to cook with it

Edible: Yes, all of it can be eaten.

The leaves and the bulb are used in cooking for their oniony taste. Leeks are commonly used in potato leek soup.

POTATO LEEK SOUP

Time: 30 minutes

Serving Size: 4 servings

Prep Time: 5 minutes

Cook Time: 25 minutes

Ingredients:

- 4 cups Ramps
- 4 cups red potatoes
- 4 cups broth
- 1 cup heavy cream
- salt and pepper to taste
- optional topping: bacon and sour cream
- flour for thickening

Instructions:

1. Clean and chop whole leeks and potatoes.
2. In a pan, cook bacon and use grease (if opting out, use butter).
3. Add leeks and potatoes, so leeks are soft.
4. Mix in a tbsp of flour.
5. Add broth, bring to a boil, and reduce to simmer.
6. Cook until potatoes are soft.

7. Add cream. Don't let it boil.
8. Remove from heat, and strain some, all, or none of the vegetable to blend into a puree, based on preference.
9. Serve hot or cold, top with bacon and sour cream.

Author: Leeks are considered a delicacy, but it is important to only take what is needed. If there aren't many leeks in the areas, leave them so that they can repopulate.

88

RED CLOVER

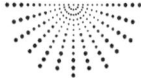

*R*ed clover is scientifically known as *Trifolium pratense*. There are a few common clovers that we see, the others being <u>wood sorrel</u> and <u>sweet clover</u>. There is a white and yellow sweet clover, not to be confused with white clover. While clover more closely resembles the red clover in flower, sweet white clover being a long spike not global foret.

IDENTIFICATION OF THE PLANT

The young plant is only about an inch tall, with three round leaves and a slightly white splat of white in the middle of each green leaf.

As the plant matures, it grows a flower stalk. The three pattern leaflet alternates up the stem. The floret is global, being about half an inch. The individual flowers are small and narrow, but long tubes that jut out from a center looks spikey. They are white and tinged with a purple/pink at the ends.

Where to gather it

The red clover is familiar to most people. Most people

know the small three leaf clover that makes its way into almost every lawn. It likes moist, shady areas. Blooms mid spring to early summer.

How to gather it

Most lawns are trimmed, which means that the flower does not get a chance to grow, so any grassy, overgrown areas are a good place to look. Pluck flowers when in bloom or leaves anytime.

How to cook with it

Edible: The flowers and leaves can be dried or used fresh. They are diverse and can be used in salads, soups, and tea. Root can be eaten raw or cooked, although cooked is not prefered.

89

ROSE

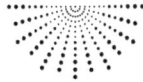

The rose is a genus of many flowers with a large diversity, the genus is scientifically known as *Rosa*. To keep it simple, we will be talking about wild roses. These roses are quite different from the ornamental roses that we typically know. Wild roses are also diverse, they are not a separate genus but a group of roses in a subcategory. *R. acicularis* is commonly referred to as the wildrose but so is *R. canina*, or the dog rose, or even *R. palustris*, the swamp rose, and so on.

The diversity of wild roses should not overwhelm the forager though. They are all fairly similar looking and can be used in the same way, there can be more preference for one than the others. If you are using roses from a flower shop, remember that like produce, there can be things sprayed on it that you are not aware of. Many flower shops and farms use pesticides, dyes, and other preservatives. If you are going to reuse flowers that were given to you, ask where they are from and see if the florist is able to tell you if it is organic or safe to use in food.

. . .

IDENTIFICATION OF THE PLANT

The wild rose is typically grown as a shrub, some getting to almost 10 feet tall, with some even staying shorter, only reaching 2 or 3 feet. These shrubs can also be thick, lush with foliage, or be more spindly with sparse foliage. The leaves tend to be an inch or so long. They are round but come to a subtle tip. They are typically more watery and a medium bluish green. The edges are slightly serrated. The stem is smooth and green when young, but becomes more woody as it gets older, covered in short thorns.

The flower ranges from magenta pink to a white at about five inches wide. They typically have five that are in full bloom, splaying out completely into a star shape. The petals are very delicate. The center is about a half inch wide, pale to golden yellow, and has a few rings of short stigma.

Rosehips are round, sometimes more flattened into a disk, or elongated like an olive (about the same size). They are an orangy red and get transparent and wrinkly over the winter. They are filled with seeds and a little fluff. Like a blueberry, it has a crown, but on the bottom. This crown is about five long (up to an inch) and narrow.

Where to gather it

Wild roses can vary depending on their preference of landscape. Some prefer coastlines, others lowlands or even sandy, well drained places. This means that no matter where you are, you will probably have luck in finding a wild rose. The flower blooms in early summer. The rosehip starts to form in late summer, so it is best to harvest in fall, after the first frost is possible. They should be reddish orange and firm, but a skin should have formed.

How to gather it

Wild roses bloom in the early summer, and it is usually for a short period of a week or so. Since they are not dense like ornamental roses, the petals fall off and fall apart really easily. If you can, harvest *only* the petals, leaving the center and bracket. This will allow the rosehip to form as well, so that you can harvest both the flower petals and the fruit from the same flower. Of course, be careful because roses have thorns.

How to cook with it

Edible: Yes, the petals and the fruit.

The rose is a super diverse flower, and it has a lot of potential in the kitchen. Its petals can be made into rose tea, rose water, distilled rose water, a garnish, and so on. The rosehips are also super useful, typically made into a jam.

DRIED ROSEHIPS

Time: 24 hours
Prep Time: 24 hours
Cook Time: 0 minutes
Ingredients:

- 1 cup rosehips

Instructions:

1. Preheat the oven or preferably a dehydrator to as low as possible, about 110 °F.
2. Wash the rosehip and dry them well.
3. Cut off the crown and other pieces attached.
4. Lay them out evenly on a tray, try not to overcrowd them.
5. Put them in the oven and let them dry.
6. This might take a while, sometimes over 24 hours. Choose a time when this can be done safely.
7. When they are completely dried, they should look like raisins. For best use, they can now be stored

for a year. You can store them whole or grind them.

8. Use for spice, teas, and more. They can be used fresh, but this way they can be stored for longer for use until next year.

Author: The rose is such an exciting flower to learn about when to foraging. It is so well known for its beauty, it is often forgotten when it comes to usefulness. It is used for beauty, fragrance, medicinal use, and food. They are full of antioxidants and vitamins. More than a pretty face.

90

SASSAFRAS

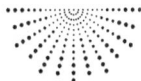

*S*assafras is a genus of trees, but it is very small. In the Americas, the *S. albumin* is the scientific name that can also be called the silk, white, and red sassafras. There is some debate about the use of sassafras because it has been linked to cancer causing chemicals. The FDA has banned the use of this plant, more specifically the root and bark. Sassafras is also used as a drug that gives people a euphoric feeling. This plant should be used at your own risk, although many people claim use in moderation is probably safe.

IDENTIFICATION OF THE PLANT

The white sassafras grows up to 70 feet tall. The bark of the tree is an ashy brown with deep grooves. The trunk tends not to be as thick as some other trees. The leaves are distinct with three lobes. The middle lobe is slightly longer and rounded. The side lobes are more narrow, sometimes giving the leaf an arrow look (pointing at the base). The edges are smooth. The leaf is a little thicker and slightly glossy. Leaves are alternate.

The flowers are yellow. They branch off in small clusters at the end of branches. The petals are paler with more of a green hint to them. They fold back into a star shape but with six narrow, pointed petals. The stamen are more noticeable, being long with large stigma at the end. The fruit is a small black berry on a bright red bracket and stem.

The sapling might have the lobed leaves or they might have simple shaped leaves. Sometimes two lobes are a morph of the two.

Where to gather it

The plant can be found in woodland areas, especially near the edges of forests or in fields. It likes well-draining sandy soil. Gather shoots in the spring. Harvest root in the fall. Leaves and twigs in spring.

How to gather it

Harvest by finding full grown sassafras trees. In this area, there should be smaller trees (saplings) that you can pull for the root.

How to cook with it

Edible: Yes.

Twigs and leaves are edible raw or cooked. They can be dried and used as a spice. The roots are known for making root beer.

HOMEMADE ROOT BEER

Time: 53 minutes

Serving Size: 10 servings

Prep Time: 8 minutes

Cook Time: 45 minutes

Ingredients:

- 1 cup roots
- 4 cups water
- 2 cloves
- 1/2 tsp fennel
- 4 allspice berries
- 1 small stick cinnamon
- 1/4 cup molasses
- 1 cup sugar
- 2 quarts soda water

Instructions:

1. Clean and cut the roots into pieces.
2. In a pot, bring water, roots, and spices to a boil.
3. Turn down heat and simmer for half an hour.
4. Add molasses and simmer for 5 minutes.
5. Strain.
6. Add back in the pot and add sugar.
7. Simmer for 5 minutes.
8. Let cool.
9. Add carbonated water 1:2.

9 1

SHEPHERD'S PURSE

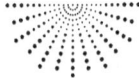

*S*hepherd's purse is a herbaceous flowering plant that goes by the scientific name *Capsella bursa-pastoris*. It is used in many Asian cuisines. It is also used as a medicine to treat a multitude of things like infection and health problems like circulation. This plant should not be consumed while pregnant.

IDENTIFICATION OF THE PLANT

The plant grows to about a foot tall. It grows a crown base of leaves that are only about 3-4 inches long. Some of the leaves might be more smooth on the edges, while some are more sharply lobed. The plant then grows a stem with blade-like leaves that are about one inch long. The plant may branch a little.

The top of the stem is the flower, the floret is more dense at the top and becomes less dense as it moves down the stem. The individual flower is very tiny star-shaped with four petals. The flowers turn into triangular shaped, flat seed pods.

Where to gather it

The plant can be found in places where the soil is soft but with a higher moisture content than sand. It also likes full sun.

How to gather it

Harvest the roots in the fall. Harvest the leaves while they still look fresh and green.

How to cook with it

Edible: The leaves are best when fresh. The plant is used for its peppery flavor. The root can be dried and ground to use instead of ginger.

SHEPHERD'S PURSE DUMPLINGS

Time: 25 minutes

Prep Time: 15 minutes

Cook Time: 10 minutes

Ingredients:

- 1 cup Shepards purse
- 1 cup minced meat
- 1 onion
- 2 tsps soy sauce
- 1 tsp vinegar
- dumpling wrappers

Instructions:

1. Clean and chop the shepherd's purse.
2. Add it to a bowl with the meat, chopped onion, soy sauce, vinegar, and spices.
3. Mix.

4. Take a tsp of the mix and add it to the center of the dumpling wrapper. Fold it in half and pinch the edges closed (adding water can help make it stick).
5. Once the dumplings are made, boil a pot of water.
6. Add the dumplings and cook for 10 minutes.
7. Serve them, freeze them, or add some oil in a pan, heat and fry the dumplings until golden brown
8. Serve with rice vinegar or soy sauce.

SILVERBERRY AND AUTUMN OLIVE

The silverberry and the autumn olive are similar, although the silverberry is more rare. The autumn olive also goes by the scientific name *Elaeagnus umbellata* and many more common names. It has become invasive in the US.

IDENTIFICATION OF THE PLANT

The autumn olive is a short tree or shrub that grows up to 10 feet tall. The branches are covered in longer thorns. The shrub is denser looking and has leaves that are two inches long, very thick, and semi-glossy that grow in an alternate pattern. They have smooth edges, and the bottom of them are paler than the cool toned green top.

The small flowers grow in small clusters in uniform rows off the branches. The petals are white and star-shaped with four petals and a tubular center. The petals are long, narrow, and end at a point.The berries grow in small clusters that droop. They are a little smaller than a marble and transparent red. When young, they are yellow.

Where to gather it

The plant can be found in disturbed areas and are commonly found on the side of the road. It's best not to harvest from near roads, but think of some similar areas you can reach.

How to gather it

Harvest the berries when they go completely red in the fall.

How to cook with it

Edible: Yes, the berries are edible.

Autumn Olive Jelly

Time: 30 minutes
Prep Time: 5 minutes
Cook Time: 25 minutes
Ingredients:

- 2 cups Autumn olive
- 2 cups water
- 1 cup sugar
- 2 tsps lemon

Instructions:

1. Bring water and fruit to a boil.
2. Cook for 10 minutes.
3. Strain with a large hole strainer and squeeze as much juice as possible out of flesh, seeds, and other impurities.
4. Add sugar and lemon to the juice in a clean pot.
5. Bring to a light boil, reduce heat slightly, and keep a steady light boil for about 10 minutes.
6. Add jelly to jars, seal the lids, and label.

93

SOLOMON'S SEAL

*olomon's Seal is a genus of plants known as *Polygonatum*. Great Solomon's seal, or *P. biflorum*, is native to eastern US. This plant has many look-alikes, see chapter 39 on false Solomon's seal.

IDENTIFICATION OF THE PLANT

Soloman's seal has a single stem that is semi-erect, slightly drooping, and stands 3-4 feet tall usually. The leaves alternate in two parallel rows. The leaves are simple with parallel veins. Their leaves are a few inches long and slightly elongated to a point.

The flowers droop in small clusters on short stems. The flowers are white, bell shaped, but elongated, not round, and slightly scalloped. The berries replace the flowers, ebing small dark round and drooping in small clusters. The root is a long rhizome.

Where to gather it

It prefers shaded areas with well draining soil. Typically

flowers in May. I would recommend harvesting young shoots in March or April.

How to gather it

Harvest shoots by cutting the young, tender plant before it matures. Harvest roots whenever.

How to cook with it

Edible: The young shoots can be cooked like asparagus. Root is used as starch.

Soloman's Seal Root

Time: 45 minutes

Serving Size: 4 servings

Prep Time: 5 minutes

Cook Time: 40 minutes

Ingredients:

- 2 cups solomon's seal root
- 2 tbsps butter
- 3 pots water
- 4 wild leeks
- pinch of salt and pepper for taste

Instructions:

1. Clean and chop the roots into 1 inch pieces.
2. Bring water to a boil with the roots.
3. Strain the water out of the pot once boiled.
4. Refill with water and boil again.
5. Repeat steps 3 and 4 one more time.
6. Strain water one more time.
7. In a pan, heat butter and chopped leeks.
8. Add roots, salt, and pepper, and cook on medium high for 5 minutes.

SOW THISTLE

*S*ow thistle is a name for a group of flowers. However, in this chapter we are talking about common sow thistle or *Sonchus oleraceus*. It might also be called milk thistle or soft thistle, among other things. It is part of the same family of dandelions, but they are different.

IDENTIFICATION OF THE PLANT

The Sow thistle, with string like petals that are a dark yellow, looks very similar to the dandelion. The main difference is that the sow thistle branches and has multiple flowerheads. Before they bloom, they make a teardrop bud.

The leaves come from a center point at the base of the flower's stem. These leaves are up to five inches long, narrow with irregular reverse scalloped pattern, and a jagged edge. The sow thistles leaf end is round. The leaves are lined with tiny pickles. The whole plant can look a little more hardy than the herbaceous dandelion.

Where to gather it

The sow thistle is a weed and likes disturbed areas like

the side of the road. Often found in lawns or near fences. They like moist soil.

How to gather it

Harvest stems and leaves by cutting the young, tender plant before it matures.

How to cook with it

Edible: The young leaves and stems can be eaten, but mature plants could be poisonous. They are used similar to spinach, raw or cooked. It is a good food to forage because it is high in antioxidants and other minerals.

STAGHORN SUMAC

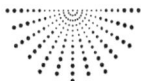

The staghorn sumac is scientifically known as *Rhus typhina*. See chapter 41 for fragrant sumac. Poisonous look alike is poison sumac that is part of the poison ivy family. Poison sumac is typically short, does not grow red fruit, with short leaves that don't put on bright red show in fall like the staghorn.

IDENTIFICATION OF THE PLANT

Staghorn sumac is a shrub that grows to almost 20 feet tall. It has a light brown/gray stem that is covered in a soft orange fuzz. The leaves are compound with leaflets reaching 20 inches. Individual leaves are narrow, five inches long, and smooth edged. The leaves are not dense foliage and the bottom few feet of trunk/ stem is easily accessible. It is bright orange red in fall.

Flowers are noticeable as they are green and small. They form the shape of a spike wherever the fruit appears. Fruit is bright red, small, fuzzy balls that grow in triangles spike up to one foot long.

Where to gather it

Staghorn sumac can be a little invasive. They like infertile soil with good drainage like sand. The fruit is ready mid to late summer and stays on the shrub throughout the winter.

How to gather it

Rub the berry in between your fingers and taste the juice. It should taste tart. Best harvested when dry. Trim with cutters underneath the spike.

How to cook with it

Edible: The berries are often dried and ground to be used as spice in savory dishes. Adds a tart flavor. Also used to make tart drinks.

STAGHORN SUMAC LEMONADE

Time: 2 hours, 30 minutes
Prep Time: 2 hours
Cook Time: 30 minutes
Ingredients:

- 1 liter sparkling water
- 2 cups water
- 2 cups sugar
- 5 tbsps sumac

Instructions:

1. In a pot, bring water and sugar to a boil.
2. Turn down the heat until sugar is dissolved and water has reduced by about half.
3. Let cool until warm.
4. Add 5 tbsps fresh sumac, cover and set in a sunny window for a few hours.

5. In a glass or jug, add sparkling water, some syrup. Taste and add more if needed.
6. Don't over stir the carbonation.
7. Can be served with ice and garnished with a lemon slice.

Author: Sumac berries are so strange that it can be hard to believe that they are edible. Sumac can be one of the more exciting foraging plants for beginners because it is strange enough and fairly easy to identify.

THISTLE

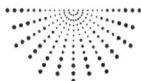

*T*histle is a common name for a group of plants that have thistles. *Cirsium edule*, edible thistle, is the plant that will be discussed in this chapter.

The burrs that this plant creates is similar to many plants that have burrs, including the burdock. Handling these plants can be dangerous because of the many look-alikes, as well as the sharp burrs.

IDENTIFICATION OF THE PLANT

When the plant is young, it is just a crown base, which is a centralized spot in the ground where the leaves fan out from. When they are at this stage, the leaves tend to be less lobed and jagged, but still reach up to six inches long.

It can grow seven feet tall with a red, hardy spine keeping it erect. It starts to branch and the leaves, including its point, can get bigger and more narrow. The stem is erect, herbaceous, and covered in thorns. The leaves are also lined in thorns.

Known for their burrs, which contain their tiny

purple/pink flowers at the top with long, white stamen. The burrs are green and brown when dried. They are covered in what looks like one inch thorns but have a small hook on the end for clinging on to passersby.

Where to gather it

This is a common weed that is found in ditches and abandoned places.

How to gather it

Harvest stems and leaves by cutting the young, tender plant before it matures. Harvest the root in the fall. Make sure to wear gloves and cut off all spokes before consuming.

How to cook with it

Edible: Yes, all of it can be eaten.

THISTLE SOUP

Time: 35 minutes
Serving Size: 4 servings
Prep Time: 5 minutes
Cook Time: 30 minutes
Ingredients:

- 1 cup potato
- 4 cups broth
- 1 onion
- 2 cloves garlic
- 1 cup thistle root
- 1 pinch salt for taste

Instructions:

1. Clean and chop the onion, root, and potatoes.
2. In a pan, heat butter.
3. Add onions and garlic.

4. Add ⅓ cup broth, potatoes, and the root. Cover until soft.
5. Transfer to pot.
6. Add broth, bring to boil, and reduce to simmer.
7. Cook until potatoes and root are soft.
8. Add cream. Don't let it boil.
9. Remove from heat and strain some, all, or none of the vegetable to blend into a puree based on your preference.

VIOLET

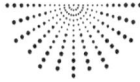

$\mathcal{T}$he term 'violet' is a genus of flowers known by their latin name *viola*. There are a lot of violets that grow in the Northeast. One of the more common violets is the *V. sororia*, or the common blue violet. These flowers have been used medicinally, mostly due to their vitamin A and C content.

IDENTIFICATION OF THE PLANT

These common blue violets don't get more than a few inches off the ground. The leaves are heart-shaped. The base of the leaf curls out hiding the stem connecting in the middle, creating a cupping effect. They are serrated on the edges and textured like wrinkles in shape.

The flowers are star shaped with oval shaped petals. The petals are typically a purple to blue color. The veining on the petals shows up a little darker. The center of the flower has a white and dark purple pattern and a white hairy looking stamen.

Where to gather it

The plant can be found in areas where the soil retains more water, but not wetlands. They like shade.

How to gather it

Harvest shoots the blooms in the spring. The flowers are so small that to get a usable haul you need to find a patch with a decent amount of violets.

How to cook with it

Edible: It can be used in salads, as a garnish, or infused in vinegar.

VIOLET SYRUP

Time: 20 minutes
Prep Time: 10 minutes
Cook Time: 10 minutes
Ingredients:

- 1 cup Violet blooms
- 1 cup water
- 1 cup sugar

Instructions:

1. Clean the flowers.
2. Bring water and sugar to a boil.
3. Boil the simple syrup until reduced by half.
4. Take off heat and add the violets (not when boiling).
5. Strain out the violets after 5-7 minutes.
6. You can seal this jar or store it in the fridge.
7. Mix with vinegar to make salad vinaigrette.
8. Mix with lemon and water to make juice.

98

WATERCRESS

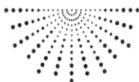

*W*atercress is known by its scientific name *Nasturtium officinale*. It is related to the nasturtium, but unlike the regular garden flower, this plant is aquatic. There are two look-alikes that you should look out for when foraging for the watercress: fools watercress and lesser water parsnip. Fool's watercress smells like carrots, while watercress does not. The difference between the lesser water parsnip is more important though because it is poisonous. Lesser water parsnip does grow a flower, but the other way to tell between the two is the rings on its stem.

IDENTIFICATION OF THE PLANT

Watercress can be identified best when it's in water. The plant is simple, a thin herbaceous stem, and round, cupping, smooth, almost succulent-like leaves. The edges of the leaves are slightly scalloped. They might be in the water completely or reaching a few inches above the water.

The flower of the watercress is a small florret with a tiny

star shape, four petal flowers. The center buds of the flower clusters are usually not budding.

Where to gather it

The plant can be found in shallow running water.

How to gather it

Harvest shoots by taking scissors and cutting a patch of watercress. Can be used from spring to fall. Make sure to wear your rubber boots and have someone with you when foraging near water.

How to cook with it

Edible: The leaves, stems, and flowers are used. Everything can be eaten raw or cooked. Has a peppery flavor.

WATERCRESS SALMON

Time: 25 minutes

Serving Size: 2 servings

Prep Time: 5 minutes

Cook Time: 20 minutes

Ingredients:

- 1 cup watercress
- 2 salmon filets
- 1 tbsp butter
- 1 tsp honey for taste
- 1 pinch salt, pepper, dill, and lemon

Instructions:

1. In a pan on medium high, heat up butter and spices.
2. When hot, use tongs to add salmon to the pan.
3. When one side is cooked, flip the salmon.

4. Add the watercress and brush the honey on the salmon.
5. Cover with a lid.
6. Turn down the heat, and when the watercress has cooked down it should be done.
7. Serve with a slice of lemon.

99

WILD BEAN

The wild bean, also called wild kidney bean or *Phaseolus polystachios*, are the only members of its genus. The wild bean can not be eaten raw like the kidney bean.

IDENTIFICATION OF THE PLANT

If the plant is crawling on the ground, it might be hard to notice because of the thin stem and understated leaves. The vine grows up to 1o feet. The leaves are opposite on the stem. The leaves are actually leaflets of three basic shaped leaves, but they can overlap and look like one lobed leaf.

The pea flower is unusual as it is not symmetrical. The upper side of the small hanging flowers are two short, cupped petals that are a lighter pink and create a hood. The bottom two petals do not touch, jutting out opposite angles. They are longer and more narrow, as well as darker pink in color. The center is a purplish tongue that sticks out slightly with a pearly looking stigma at the end.

The beans are like green bean pods, long and thin, but

wild beans tend to be a little thicker and shorter. They hang off of the stem by itself or in rows. The beans get a dark green to brown color and patchy. They get harder and thicker as they dry.

Where to gather it

The plant can be found in dry areas with sand. They like the edges of forests and thickets. This is a vine that likes to climb, not as much as crawl. It is harvested in the summer.

How to gather it

Harvest shoots and the seed pods when they are green. They should be firm, but don't let them age too long to the point the seeds are bulging out of the pod. You can let the bean dry in the pod and treat it like a dried kidney bean.

How to cook with it

Edible: Yes, the bean is edible.

Unlike green beans, this bean must be cooked to make it edible.

WILD BEAN AND RICE

Serving Size: 4 servings

Ingredients:

- 1 cup dried wild beans
- 2 cups water
- 1 tbsp butter
- spices: garlic, onion, cumin, chili powder, salt, and black pepper
- 3 cups cooked rice

Instructions:

1. Fill a bowl or a pot with cold water and add the beans.

2. Let it soak overnight. The other option is to boil the beans, change the water, and cook until they are soft.
3. Drain the water and add in the rice and spices with the butter. Optional: add some tomato sauce.

100

WILD CARROT, QUEEN ANNE'S LACE

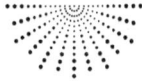

*W*ild carrot is known scientifically as *Daucus carota*. This plant can look similar to poison hemlock. Poison hemlock has no hairs on its stem and a purple hue. The flower also tends to be small clusters of flowers that create the floret, instead of one large plate. Hogweed also looks like queen anne's lace and is poisonous. Hogweed is much bigger and has a smooth stem.

IDENTIFICATION OF THE PLANT

The wild carrot grows up to three feet tall. The stem is herbaceous but thick. The branches come out in opposite rows. The lower leaves have three lobes. They are compound, up to a foot long. The leaves on the floret are opposite rows of pairs. The leaves near the base are longer and have more points than the leaves at the top of the leaflet.

The flower is a round and flat topped floret. It consists of a few branches that end in a hollow circle of tiny white flowers. The inner flowers are smaller, the pattern making it look like lace. The flowers are about eight inches across at the most.

Where to gather it

This is a common weed and is found in ditches and abandoned places.

How to gather it

Harvest roots in the late summer to early fall by tugging the plant. The flowers can be cut off when they are in full bloom or until fruit appears.

How to cook with it

Edible: The flower, root, and fruit are all eaten for their carrot flavor. They are used in all dishes where carrots are used.

PICKLED JUNIPER BERRIES

Time: 10-15 minutes

Prep Time: 5 minutes

Cook Time: 5-10 minutes

Ingredients:

- 1 cup Queen Anne's lace root
- 1 cup pickling vinegar
- 1 tbsp salt
- 1 cup water
- 1 tbsp sugar
- spices of choice: garlic, onion, and pepper

Instructions:

1. Wash queen anne's lace, chop it and other produce.
2. In a pot, boil water, vinegar, salt and sugar until the sugar is dissolved.

3. While the brine is heating, add a mixture of spices and the chopped onion, root, and flower in sterilized jars (heat proof). Leave room at the top, but the pieces should be fully submerged.
4. Pour hot brine over the pieces in the jars.
5. Seal, label, and store in the fridge. Wait at least a week to pickle.

101

WILD GINGER

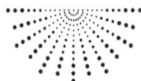

*W*ild ginger is also known as Canadian wild ginger and *Asarum canadense*. This plant has been used for medicinal purposes, mostly relieving common cold problems. This plant should not be used like ginger because it has been linked to cancer causing chemicals. That being said, that is strictly the root and not the rest of the plant.

IDENTIFICATION OF THE PLANT

Wild ginger grows in small little clusters of stems. It is about a few inches tall at the most. It might go unnoticed if it is a lone plant, but also, if clustered in a big group, it might look more like a vine or ground cover. The stems are herbaceous and green.

The leaves are the most identifiable part of the plant. It is heart shaped, but the leaf faces up and the middle of the heart bumps are quite deep, looking more like a dimple. The leaves are slightly wrinkled, with smooth edges and a medium cool green to its color. Leaves are covered in fine hair. Only one leaf per stem.

The flower is a dark burgundy red. It has three petals that are trumpet-like in shape, but the petals are very long, curling out like strings. The flower is slightly hairy.

Where to gather it

The plant can be found in woodland areas especially near banks, but it likes shady places without much sun. It likes lots of rich soil.

How to gather it

With a shovel, dig up the wild ginger when it is ready. Like most roots, it is best harvested in the late fall or (although less preferred) in the early spring.

How to cook with it

Edible: Yes, the root is used fresh or dried. Avoid eating too much of the root. Cook the root in water because the toxin is not water soluble. This tea is what is used for medicinal purposes.

WILD GINGER ICE CREAM

Time: 10 minutes

Prep Time: 10 minutes

Cook Time: 0 minutes

Ingredients:

- 2 cups heavy whipping cream
- 2 cups half and half
- 1 cup white sugar
- 2 tsps vanilla extract
- 2 tbsps wild ginger
- optional: other warming spices like nutmeg and cinnamon

Instructions:

1. Combine all of the ingredients and stir until the sugar is completely dissolved.
2. In an ice cream maker, add the mixture.
3. Turn on the ice cream maker, making sure to occasionally stir the sides.
4. Depending on your ice cream maker, don't let it get too cold because it will end up hard and not creamy.
5. Once your ice cream is done, serve or freeze in a sealed container.
6. Optional: use one of the many fruits in this book to make a syrup to go on top.

102

WILD LETTUCE

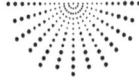

*ild lettuce is interesting because, while its latin name is *Lactuca virosa*, it is also known by many other names, such as bitter lettuce, opium lettuce, poison lettuce, and more. This plant has a latex-like sap that is used in medicine, usually used to help people sleep. It also has been used to ease pain. Because of the medicinal qualities of this plant, it is also a good idea to not over consume it. If you are allergic to latex plants like this and other plants with a latex sap, this can be dangerous to you. Wild lettuce has some look-alikes like the thistle and the dandelion, but they are also edible.

IDENTIFICATION OF THE PLANT

This plant is called prickly and tall lettuce for a reason. It can grow four feet tall with a red, hardy spine keeping it erect. The leaves are covered in prickles, especially near the margins. When the plant is young, it is just a crown base, which is a centralized spot in the ground where the leaves fan out from. When they are at this stage, the leaves tend to be less lobed and jagged, but still grow up to six inches and rounded at the ends. As the stem grows in, the leaves end up looking almost torn. They are lobed, inverse scalloped, and jagged.

The flower is very similar to the dandelion. This flower instead has a prominent local bracket, and the leafy tube-like sepals hold the flower almost closed. When it opens more, there are less petals than a dandelion that are slightly wider, and the shape of the flower is more flat. When it goes to seed, it seeds like a dandelion too, but the global pod is less dense and the fluffy ends of the seed are more brown.

Where to gather it

The wild lettuce is a weed and likes disturbed areas like the side of the road. It's best not to harvest from near roads, but think of similar areas. It blooms all summer.

How to gather it

Harvest leaves by the young, tender plant before it matures. Even regular lettuce is better when it is young and crisp. Pick the plant as it is blooming when using it for medicinal purposes.

How to cook with it

Edible: Of course wild lettuce can be eaten like regular lettuce. It is also boiled, particularly as a tea.

WILD LETTUCE TINCTURE

   Time: 72 hours, 20 minutes
   Prep Time: 72 hours, 20 minutes
   Cook Time: 0 minutes
   Ingredients:

- 1 cup wild lettuce stems and leaves
- 2 cups 40% alcohol

Instructions:

1. Wear gloves because the plant is prickly.
2. Clean and chop the plant.
3. Dry completely.
4. In a blender, add the plant and puree.
5. In a jar, add the puree and the alcohol.
6. Let soak for 3 days, shaking once a day. Keep in a cool, dark place.
7. After 3 days, use cheesecloth to strain out the plant material. Be careful not to get too much on your skin.
8. Add stained liquid to a dropper bottle.
9. Use for sleep and pain.

103

WOOD SORREL

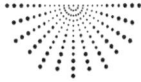

ood sorrel is a genus of flowers called *Oxalis*. One of the most common of the wood sorrels is *O. strictica*, or yellow woodsorrel, but most, if not all, wood sorrels are considered edible. They are also a part of the group of clovers that people find in their lawns. Also see the chapter on <u>red clovers</u> and the chapter on <u>sweet clovers</u>.

IDENTIFICATION OF THE PLANT

The plant is a common three leaf clover. When it is young or groomed by a lawn mower, it can stay as a one inch tall clover. This clover is a solid yellow green and lobed at the end to look like a heart. It can grow to a small patch of clovers on some spindly stems, even then staying only a few inches tall. The flower is on a spindly stem itself. It is star shaped, but the petals curl back a bit. They are bright yellow and about the size of a buttercup. The pod is green, rice shaped, with vertical grooves and slightly hair, at about half an inch long. The leaves curl up at night.

Where to gather it

Since most clovers are considered a weed in gardens, the clover is pretty common in most lawns. It blooms late spring to early fall.

How to gather it

Harvest any time it is out.

How to cook with it

Edible: Yes, all of it can be eaten, but it should only be consumed in small amounts. The leaves and flowers can be eaten raw or cooked, like leafy greens. Typically used in salads. It can also be made into teas. The fruit is tart and crisp.

WOOD SORREL SOUP

Time: 20-25 minutes

Serving Size: 4 servings

Prep Time: 5 minutes

Cook Time: 15-20 minutes

Ingredients:

- 3 tbsps butter
- 1/2 cup onions
- 4 to 6 cups wood sorrel, packed
- 4 cups vegetable stock
- 1/2 cup cream
- pinch of salt

Instructions:

1. Melt butter in a pot and add chopped onions.
2. Cook until golden.
3. On medium heat, add sorrel and salt until cooked down.
4. Take off heat and add to a blender and puree.

5. Pour back in the pot and add cream. Cook on low heat for 5 minutes, but do not boil.
6. Serve cool or warm.

Author: Since the berries can aid in digestion, although they might not be the best snack, they can be helpful to have around when in need. This plant should be considered once a forager has a little bit more experience and can feel confident in identifying a plant early in the season that has many look-alikes.

104

YARROW

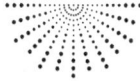

*Y*arrow is also known as sweet alyssum or scientifically known as *Lobularia maritima*.

IDENTIFICATION OF THE PLANT

Yarrow can be grown as a garden flower, although it can also be foraged for. The plant is bushy looking, having multiple stems in a dense cluster. They usually only grow a foot or two tall. The stems are thin, but dense/hard. The stem has vertical grooves. The flowers of the plant are compound. The individual leaves are very narrow like a herbaceous needle, similar to dill, they are short, and dark green. The leaflet is about six inches long, with a main margin and two rows on either side of dense flowers. It looks like a feather.

The flowers are dense firm florets. The flowers are five petal star shaped. The petals are round. The center has small craters. The shape of the floret is a roundish cluster, the face is flat, and 3-4 inches across on average. The flower is usually white or off white, but can be pink or yellow as well.

Where to gather it

Yarrow might be found blooming from the late spring through the summer. It likes direct sun, is not picky with soil, and can be found in weedy areas.

How to gather it

Use scissors because the stem is hard. Clip flowers and leaves as needed.

How to cook with it:

Edible: Yes, the flowers and leaves can be dried for spices or used fresh. Don't overcook. I'd recommended cooking it with beets and parsley flavors.

Author: Yarrow is really common in medicinal medicine; it is used in topical treatments to help healing and prevent infection.

WILD POTATO VINE

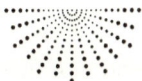

The wild potato vine, or *Ipomoea pandurata*, is also known as wild sweet potato and wild rhubarb. This seems like very different things to be compared to.

IDENTIFICATION OF THE PLANT

The wild potato vine can grow up to 30 feet long. The leaves of this plant are up to six inches long and alternate. The leaves are heart-shaped and smooth edged. Often they are tinged with bug holes.

The flower is white and trumpet-shaped. They are about two inches wide and fan out. The center tube is touched with purple. The roots might be small like a potato, but they are often large, growing up to a foot long and more carrot or yam-like in shape.

Where to gather it

The plant can be found in many places like forests, fields, and even rock areas. Since it is a vine, it likes to climb or crawl.

How to gather it

374 | NORTHEAST FORAGING FROM YOUR BACKYARD HO...

In the fall dig the plant up to get to the root, but it might be easier to take notice of it earlier in the year when it is flowering.

How to cook with it

Edible: Yes, the roots can be eaten when cooked. They are similar to a sweet potato.

Author: Like many of the root vegetables in this book, those who are foraging for a larger amount of their food intake should really take note of these kinds of plants. Many people think that low calories or low starch is a good thing, but it can be hard to find food with enough calories to sustain a healthy lifestyle. High starch foods like this one can make foraged meals heartier.

Leave a 1-Click Review!

Customer reviews

⭐⭐⭐⭐⭐ 5 out of 5

3 global ratings

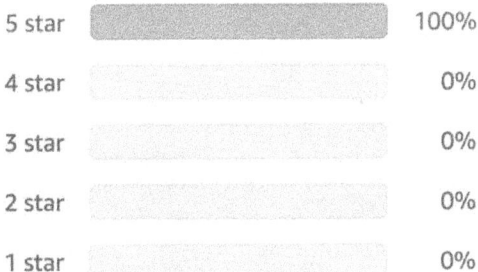

5 star	████████████	100%
4 star		0%
3 star		0%
2 star		0%
1 star		0%

⌄ How are ratings calculated?

Review this product

Share your thoughts with other customers

[Write a customer review]

AFTERWORD

This book introduces the reader to over 100 wild edible herbs, which can certainly feel like an endless amount of information. Foraging is one of those things that you can't just master by reading a book. Learning is mostly done in the field. Some people might start identifying every plant that they pick up. While it can feel endless, it is important to remember that these are the most common plants you can forage for in this area. While there are a lot of plants that are edible, you don't need to know or memorize them all. Take your time and figure out what plants work for you.

Every single plant includes a recipe for the reader. This is a good way to introduce yourself to the plant, and then it's your turn to figure out what you are going to do with it. Food has endless possibilities and, as you learn about the plant in its habitat and get to know food on a more intimate level than the sterile grocery store, your kitchen will become a place that is full of inspiration. The amazing thing about so many of these plants is that you probably already know them. Since so many are common, it means accessibility for everyone for foraging food. These plants are probably already in your backyard.

I tried to begin each chapter with a hint of an experience on the plant so that the foraging community can be shared even through this book. Whether this insight was about taste, safety, or identification, I wanted people to be able to feel more connected with the plants and feel like they already had one up to go out and look for in the forest. The dynamic relationship then becomes a whole picture—the location, the plant, the exterior knowledge, how to interact with the plant, and how to use the plant. The outdoors and your kitchen become connected too. The grocery store and your kitchen are not separate from the outdoors, they are the same.

Foraging for food is more than a single purpose. You can save money, improve your health, prevent yourself from any illnesses, spend time in nature, and have fun cooking these extraordinary recipe ideas all at the same time. It is hard to bring all of these things into a situation, but you can enrich your life by taking a step outside to see what you can use. As you focus on the idea of foraging for more, you will notice a difference in how you eat too. You can appreciate the sophis-ticated taste that comes from the variation in leafy greens. You can taste the heightened sweetness from a fresh peach. This appreciation changes what your plate looks like, and eating better food becomes a gift. These plants are also known for their different capabilities in healing. All of this can be done without having to buy into any scheme.

Food from your backyard is healthier; you know where it comes from. You can make your food become your medicine and observe yourself rising and shining. Why would you still buy vegetables and herbs if you can use the wild woods as your free supermarket? You can have more control over what goes in your body. Much of the food that we eat from the store has a list of ingredients it doesn't need. Some food gets picked before it is ripe so that by the time it hits stores it isn't rotten, but this means that the plant doesn't have time to grow the way it needs to in order to be the same tasty, healthy food.

Let's spread this knowledge together. A review can help others to gain the same insights that you just got. What feels better than helping others rise? If you enjoyed the content of this book, please take 2 minutes to leave a review on this. We also have an active Facebook group and would love to have you.

https://www.facebook.com/groups/northeasthomestead

Let's help humanity rise together!

Other Titles!

https://www.amazon.com/dp/B09PK252MF

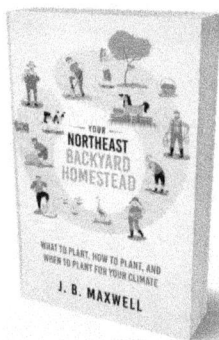

Northeast Backyard Homestead

BIBLIOGRAPHY

Benfer, A. (n.d.). *Foods indigenous to the Western Hemisphere pigweed*. AIHDP. Retrieved March 4, 2022, from https://aihd.ku.edu/foods/Pigweed.html

CDC. (2022). *Centers for Disease Control and Prevention*. U.S. Department of Health & Human Services. https://www.cdc.gov/

Colleen. (2020, December 13). *What to forage in winter: 30+ edible and medicinal plants and fungi*. Grow Forage Cook Ferment. https://www.growforagecookferment.com/what-to-forage-in-winter/

Deane, G. (n.d.). *Eat the weeds and other things, too*. Eat the Weeds and Other Things, too. Retrieved March 4, 2022, from https://www.eattheweeds.com/

Docio, A. (2017, November 20). *What is foraging? Finding your food in the wild*. British Local Food. https://britishlocalfood.com/what-is-foraging/

Ellen. (2021, December 5). *Mallow meringues: A light & airy cookie recipe*. Backyard Forager. https://backyardforager.com/mallow-meringues-recipe/

Extension: Utah State University. (n.d.). *Jame's chickweed*. Retrieved March 4, 2022, from https://extension.usu.edu/rangeplants/forbsherbaceous/JamesChickweed

Gore-Tex. (2017, November 8). *Foraging for beginners: Tips for safely gathering wild, edible foods*. https://www.gore-tex.com/blog/foraging-food-wild-plants#:~:text=Foraging%20safety

Grant, B. L. (2020, December 20). *Leaf identification - Learn about different leaf types in plants*. Gardening Know How. https://www.gardeningknowhow.com/garden-how-to/info/different-leaf-types-in-plants.htm

Haines, A. (n.d.). *Why foraging*. Arthur Haines. http://www.arthurhaines.com/why-foraging#:~:text=It%20helps%20people%20become%20more

Kanuckel, A. (2021, August 13). *The many uses for wild, edible cattails*. Farmers' Almanac. https://www.farmersalmanac.com/cooking-wild-edible-cattails-25374

King, A. (2021, June 10). *The forager's toolkit – Essential equipment and tools you need for wildcrafting* [Video]. Eatweeds. https://www.eatweeds.co.uk/toolkit

Lumen - Boundless Biology. (n.d.). *Leaves*. Retrieved March 4, 2022, from https://courses.lumenlearning.com/boundless-biology/chapter/leaves/#:~:text=Leaf%20Arrangement

Massachusetts Institute of Technology. (2019). *The Massachusetts Institute of Technology (MIT)*. https://www.mit.edu/

Meredith, L. (2014). *Northeast foraging: 120 wild and flavorful edibles from beach plums to wineberries*. Timber Press.

Mertins, B. (n.d.). *5 tips for when you need help identifying a plant*. Nature Mentoring. https://nature-mentor.com/need-help-identifying-a-plant/

Milham, A. (2013, July 10). *10 ways to use bee balm and bee balm bread recipe*. Premeditated Leftovers. https://premeditatedleftovers.com/recipes-cooking-tips/10-ways-to-use-bee-balm-and-bee-balm-bread-recipe/

MyWildflowers.com. (n.d.). *My wildflowers identification tool*. Retrieved March 4, 2022, from http://mywildflowers.com/identify.asp

Thomas S. Elias. (1982). *Edible wild plants*. Sterling Publishing Co.

Wikipedia Contributors. (2018, November 24). *Wikipedia, the free encyclopedia*. Wikipedia; Wikimedia Foundation. https://en.wikipedia.org/wiki/Main_Page

Wild Food UK. (n.d.). *Ground elder, goutweed, bishops weed, Aegopodium podagraria*. Retrieved March 4, 2022, from https://www.wildfooduk.com/edible-wild-plants/ground-elder/

WWF. (n.d.). *World Wildlife Fund Canada*. Retrieved March 4, 2022, from https://wwf.ca/?gclid= Cj0KCQiA64GRBhCZARIsAHOLriJAlUPdEVolNGqcgDUtxG2eiqIziP PCWoSPfOpFkbIHImrg5iSh888aAr63EALw_wcB

Young, D. (2017, March 20). *Ethical foraging 101: What you need to know*. LearningHerbs. https://learningherbs.com/skills/foraging/

IMAGE REFERENCES

adege. (2021, May 3). *Fiddlehead fern plant* [Image]. Pixabay. https://pixabay.com/photos/fiddlehead-fern-plant-6224434/

Alicja. (2018, December 1). *Hawthorn hawthorn fruit autumn tree* [Image]. Pixabay. https://pixabay.com/photos/hawthorn-hawthorn-fruit-autumn-tree-3843011/

angelac72. (2016, September 15). *Eastern red buds spring pink* [Image]. Pixabay. https://pixabay.com/photos/eastern-red-buds-spring-pink-1671000/

Bernell. (2016, August 10). *Staghorn sumac rhus typhina* [Image]. Pixabay. https://pixabay.com/photos/staghorn-sumac-rhus-typhina-1579339/

byrev. (2013, March 1). *Crispus curly dock* [Image]. Pixabay. https://pixabay.com/photos/crispus-curly-dock-raw-rumex-87928/

cornelnux. (2014, October 7). *Plantain grass* [Image]. Pixabay. https://pixabay.com/photos/plantain-grass-green-476851/

ignartonosbg. (2021, December 16). *Flora plant ground cherry* [Image]. Pixabay. https://pixabay.com/photos/flora-plant-ground-cherry-physalis-6872728/

JameDeMers. (2017, September 8). *Autumn olive japanese silverberry* [Image].

Pixabay. https://pixabay.com/photos/autumn-olive-japanese-silverberry-2729140/

jhenning. (2020, June 6). *Wildflower melilotus trifolia* [Image]. Pixabay. https://pixabay.com/photos/wildflower-melilotus-trifolia-5265766/

jxzh91877. (2021, August 31). *Gallant soldier flowers plant* [Image]. Pixabay. https://pixabay.com/photos/gallant-soldier-flowers-plant-6580980/

lijunwei6. (2017, April 10). *Zhujiajiao ancient town* [Image]. Pixabay. https://pixabay.com/photos/zhujiajiao-ancient-town-pawpaw-2212538/

MAKY_OREL. (2021, June 13). *Black locust robinia pseudoacacia* [Image]. Pixabay. https://pixabay.com/photos/black-locust-robinia-pseudoacacia-5291815/

Mayapujiati. (2017, September 3). *Bee amaranth flowers* [Image]. Pixabay. https://pixabay.com/photos/bee-amaranth-flowers-insect-2709238/

Nennieinszweidrei. (2021, July 7). *Thistle scraping cirsium* [Image]. Pixabay. https://pixabay.com/photos/thistle-scraping-thistle-cirsium-6391252/

Pezibear. (2018, June 15). *Yarrow plant blossom* [Image]. Pixabay. https://pixabay.com/photos/yarrow-plant-blossom-bloom-white-3474451/

siala. (2016, September 22). *Common bramble blackberry shrub* [Image]. Pixabay. https://pixabay.com/photos/common-bramble-blackberry-shrub-1687348/

SimoneVomFeld. (2021, April 16). *Garlic mustard edible seasoning* [Image]. Pixabay. https://pixabay.com/photos/garlic-mustard-edible-seasoning-5051652/

WikimediaImages. (2015, July 15). *Apiaceae hercaleum hogweed* [Image]. Pixabay. https://pixabay.com/photos/apiaceae-hercaleum-hogweed-844414

WikimediaImages. (2015, July 22). *Lamium maculatum spotted* [Image]. Pixabay. https://pixabay.com/photos/lamium-maculatum-spotted-dead-nettle-846464/

WikimediaImages. (2015, July 22). *Stellaria media common chickweed* [Image]. Pixabay. https://pixabay.com/photos/stellaria-media-common-chickweed-846435/

WikimediaImages. (2015, July 28). *Artemisia vulgaris mugwort* [Image]. Pixabay. https://pixabay.com/photos/artemisia-vulgaris-mugwort-848743/

www.ingramcontent.com/pod-product-compliance
Lightning Source LLC
Chambersburg PA
CBHW031839200326
41597CB00012B/197